Colorimetry

Edited by Ashis Kumar Samanta

Published in London, United Kingdom

IntechOpen

Supporting open minds since 2005

Colorimetry
http://dx.doi.org/10.5772/intechopen.95699
Edited by Ashis Kumar Samanta

Contributors
Sonia Karuppaiah, Manikandan Krishnan, Pubalina Samanta, Ran Aharoni, Asaf Zuck, Shai Kendler, David Peri, Murali Dadi, Mohd Yasir, Gen Motojima, Vitthal V. Vishram Chopade, Jayashri V. Chopade, Deepali Singhee, Adrija Sarkar, Md. Anowar Hossain, Ashis Kumar Samanta

First published in London, United Kingdom, 2022 by IntechOpen
IntechOpen is the global imprint of INTECHOPEN LIMITED, registered in England and Wales, registration number: 11086078, 5 Princes Gate Court, London, SW7 2QJ, United Kingdom
Printed in Croatia

British Library Cataloguing-in-Publication Data
A catalogue record for this book is available from the British Library

Additional hard and PDF copies can be obtained from orders@intechopen.com

Colorimetry
Edited by Ashis Kumar Samanta
p. cm.
Print ISBN 978-1-83962-940-2
Online ISBN 978-1-83962-941-9
eBook (PDF) ISBN 978-1-83962-942-6

For EU product safety concerns:
IN TECH d.o.o., Prolaz Marije Krucifikse Kozulić 3, 51000 Rijeka, Croatia, info@intechopen.com or visit our website at intechopen.com.

Meet the editor

Dr. Ashis Kumar Samanta is presently a professor and former Head of Department of the Department of Jute and Fibre Technology, University of Calcutta, India. He was previously the chairman of the Bureau of Indian Standards (BIS) TXD-07 Committee for Speciality Chemicals and Dyestuff; chairman, Board of Studies (BOS), Bachelor of Fashion and Apparel Design, University of Calcutta; and a regional board member for South East Asia for the American Association of Textile Chemists and Colorists (AATCC), USA. He was a Professor In Charge at the Institute of Jute Technology (IJT), India. From 2009 to 2011, dr. Samanta was a member of the Governing Council of IJT for three terms. He has also been a member of the Research Advisory Committee of the Indian Council of Agricultural Research, National Institute of Research on Jute and Allied Fibre Technology (ICAR-NIRJAFT); an expert member of the Technical Advisory Committee for Khadi and Textiles at the Mahatma Gandhi Institute for Rural Industrialization (MGIRI); member of BOS for Bachelor of Design (Textiles), Viswabharati; member of the Ph.D. Committee in Textile Technology, University of Calcutta; and the former chairman/convener and co-chairman of the Textile Engineering Division, WBSC of IE(I). He served as the assistant director of the Manmade Textiles Research Association (MANTRA), Surat, from 1997 to 1998. He is an expert member of WG-31 on Natural materials for Textiles under ISO-TC-38 and attended the plenary meetings of ISO-TC-38. On behalf of BIS, Dr. Samanta collaborated with other experts to develop ten BIS Test Standards and five ISO Test Standards for natural dyes. He has run and completed sixteen research-and-development projects and two HRD projects. Dr. Samanta is the recipient of numerous awards, including the Bharat Nirman Award in Jute and Textile Technology Education and Research in 2014; Eminent Engineer Award in 2019; Triguna Charan Sen Award in 1996 and 2014; Subject Paper Award in 2006; and a Special Merit Award in 2007, all from the Institution of Engineers (India) IE(I). He is the reviewer of many international journals including Textile Research Journal, Journal of Natural Fibres, Journal of Industrial Textiles, Journal of Textile Institute, Indian Journal of Fibre and Textile Research, Journal of Fibres and Polymers, Coloration Technology, Journal of Cleaner Production, Journal of BioResource and Technology and Indian Journal of Natural Fibres. He is an editorial board member for five national and international journals. To date, he has mentored twelve Ph.D. students and is currently mentoring three more. His main areas of research are chemical processing and finishing including natural dyes, nanomaterials, and phase change materials. He has more than 300 publications, including 135 research papers, 55 technical articles, 108 conference papers, 10 book chapters, 11 training manuals, and 4 books, to his credit.

Contents

Preface

Colorimetry is the science of quantitatively defining, measuring, and communicating a colour of any solid, liquid, or semisolid materials, instead of subjectively describing it, for producing the same colour assured by instrumental testing.

Colour is also one of the most sensitive factors determining the psychological and physical buying decisions of consumers. Hence colour mixing and matching to obtain a desirable shade of solid or liquid products is essential. Initial simple colorimetric measurements have been replaced by spectrophotometric (both absorption spectrophotometry and reflectance spectrophotometry) analysis of colours and updated with digital image analysis of colours in different sectors. Moreover, for more commercial exploitation, every industry understands the importance of colour measurement, communications, and precision matching of colours. As such, this book presents a comprehensive overview of colorimetry. It includes the following ten chapters:

Chapter 1: "Colorimetric Evaluations and Characterization of Natural and Synthetic Dyes/Pigments and Dyed Textiles and Related Products"

Chapter 2: "UV-Visible Spectroscopy for Colorimetric Applications"

Chapter 3: "Reflectance Spectra Analysis Algorithms for the Characterization of Deposits and Condensed Traces on Surfaces"

Chapter 4:"Spectroscopy and Spectrophotometry: Principles and Applications for Colorimetric and Related Other Analysis"

Chapter 5: "Basic Principles of Colour Measurement and Colour Matching of Textiles and Apparels"

Chapter 6: "Colorimetric Measurement and Functional Analysis of Selective Natural Colorants Applicable for Food and Textile Products"

Chapter 7: "A Digital Image-Based Colorimetric Technique Use for Quantification of Green Active Pharmaceuticals Obtained from Natural Sources"

Chapter 8: "Colorimetry in Nuclear Fusion Research"

Chapter 9: "Evaluation of Camouflage Coloration of Polyamide-6,6 Fabric by Comparing Simultaneous Spectrum in Visible and Near-Infrared Region for Defense Applications"

This book presents many scientific case studies on practical guidelines as well as difficulties and solutions for colorimetric analysis in the textile, pharmaceutical, paint, food, defence, nuclear fusion, and other relevant industries. It is a useful resource for dyers, colourists, scientists, researchers, and students engaged in colour analysis, colour mixing, and colour matching.

As editor of this book, I wish to acknowledge the constant support of all the administrative executives, author service managers, and the editorial and production team at IntechOpen, without whom this book project would not have been possible. I am also thankful to the contributing authors for their excellent chapters.

Ashis Kumar Samanta
Professor,
Department of Jute and Fibre Technology,
University of Calcutta,
Kolkata, West Bengal, India

Section 1

Defination, Measurement and Evaluation of Colours

Chapter 1

Colorimetric Evaluations and Characterization of Natural and Synthetic Dyes/Pigments and Dyed Textiles and Related Products

Ashis Kumar Samanta

Abstract

This book chapter covers principles and few case studies on colorimetric Estimation of (i) determining purity/active ingredient % of selective dyes/pigments (ii) Identification of any colorants to distinguish from other similar compound, (iii) Measurement of surface colour strength of a dyed textile, (iv) Measurement of colour differences by estimating DE, DL*, Da*, Db*, DC and DH values, (v)-Computer-aided colour match prediction for any standard shades, (vi) Estimation of compatibility of two dyes/colourants to use for compound shades, (vii) Determination of rate of dyeing, dyeing isotherm and dyeing kinetics to control dyeing, (viii) Optimization of dyeing process variables, (ix) Precession grading of Colour Fastness of dyed textiles on fading under different ways/agencies and (x) Estimation of Soil Removal efficacy of different detergent used for textiles. These colorimetric measurements are found to be very useful for effective process and product control of dyed textile materials. Selected Case studies on all the above colorimetric applications with specific example or experimented data are discussed for each of the method under reference. Finally, the other applications of colorimetric analysis besides textiles industry are also mentioned in concluding remarks.

Keywords: colour quantification, CIE-L* a* b* colour space, colour matching of textiles, standardization of dyeing process variables, colour fastness grading, UV VIS absorbance spectrophotometer, UV VIS reflectance spectrophotometer, soil removal efficacy of detergent

1. Introduction

UV VIS Absorption spectrophotometer (applicable for coloured liquid) and UV VIS reflectance spectrophotometer (applicable for flat samples of coloured solids) are the two major equipment now being used in colorimetric evaluation related labs for textiles and other industries. Colour measurement of liquid dye solution or colorimetric titrations is known with the advent of UV Vis Absorption spectrophotometer. Colour measurement of solid substances was quantified by CIE internationally in 1923, which was further revised in 1976 and is continuing [1–4]. Colour matching theory was made commercially applicable in 1950s. Till then, so many varied applications of colorimetric evaluations have made precision process control and product control possible for coloured textiles.

2. Principles of colorimetric/spectrophotometric evaluation of coloured substance using UV: VIS absorbance spectrophotometer

Beer's Law states that the amount of light absorbed is directly proportional to the concentration of the colored solute in the solution.

$$\mathrm{Log}_{10}\frac{I_0}{I_t} = \varepsilon c \qquad \text{(i)}$$

where, ε = proportionality constant and c = concentration of the solute in solution.

Lambert's Law states that the amount of light absorbed is directly proportional to the length and thickness of the solution (thorough which light is passed through) under analysis.

$$\mathrm{Log}_{10}\frac{I_0}{I_t} = \varepsilon b \qquad \text{(ii)}$$

where ε = proportionality constant and b = length/thickness of the quartz cell in which solution is tested.

So, combining these two laws, called Beer-Lambert Law [3, 4]:

$$A(\text{Absorbivity or Absorbance}) = \mathrm{Log}_{10}\frac{I_0}{I_t} = \varepsilon bc \qquad \text{(iii)}$$

where I_0 = intensity of the incident light, It = intensity of the transmitted light, c = the concentration of absorbing substance/solute in the solution, b = the path length/the distance the light is passing through the absorbing solution, and A = absorptivity/Absorbance and ε = proportionality constant dependent upon the absorbing substance, the wavelength of light used, and the units used to specify c and b.

In simple colorimetry, the entire visible spectrum (white light) is used to pass through the solution and consequently the complementary colour of the one absorbed, is observed as transmitted light. In UV VIS absorbance Spectrophotometer, a monochromatic light or a narrow band of light radiation is used, replaced the colorimeter and then this instrument is called Absorbance spectrophotometer or reflectance spectrophotometer, differentiating by measurement parameter i.e. measuring as absorbency or optical density of transmitted light intensity for colored solution.

Limitations and Cares for measuring absorbance/optical density parameter of liquids:

i. Beer-lambert law does not hold good for a concentrated solution. So sufficient dilution is necessary to obtain correct and reproducible results. Dilution to 50 to 100 times is preferably used.

ii. Beer-lambert law does not hold good, if the solute/coloured liquid under measurement, has ionizing, dissociation or aggregation/association tendency or complex-forming tendency in the solution.

Example: Benzoic acid in Benzene solvent form dimer, i.e., aggregates as dimer, and Potassium dichromate on higher dilution, dichromate ions are ionizing or dissociating into chromate ions, which are the causes of deviation of correct reading in both these two cases.

i. If the colour liquid has fading tendency with time, then the sample is faded away due to instability of coloured molecules/solutes and hence incorrect results are obtained.

ii. Presence of any impurities like fibre dust, residual dye bath additives/ electrolyte etc. (comes to the coloured liquid solution during extraction of coloured substance/dyes/pigments from a dyed textiles or during measuring residual coloured liquor of exhausted dye bath effluent) and causes incorrect result.

iii. Use of electrolyte at higher concentration usually shift the λmax values and changes the extinction coefficient/coefficient of absorption etc. and hence occur deviation in results.

iv. Presence of any additives changes/makes alterations in the refractive index values of the coloured solution and hence it gives wrong results.

v. Changes in pH of solute/coloured liquid causes deviation in results.

While for Solid coloured samples, surface reflectance values are measured for any solid-coloured substance, the measurement parameter is reflectance (R values at different wavelengths user-chosen wavelength or preferably at maximum absorbance wavelength, i.e., λmax) and the instrument used for R values of solid coloured substance, is called UV-VIS Reflectance Spectro-photometer.

Thus, when it is required to measure colour from a solid dyed/printed surface, the measurement parameter is not absorbancevalues, but is Reflectance values (R), i.e., reflected light intensity from a solid surface of dyed textiles/coloured/coated polymeric film/plastics etc. R_L – Rs UV-VIS reflectance spectrophotometer is used having different viewing angle with setting facility of measuring specular reflectance or diffused reflectance, with UV-in (On) and out (Off), with large viewing angle or small viewing angle, with D65 or other standard illuminant light ambience etc. having options of changed testing parameters. Both these spectrophotometers are not limited to the visible spectrum only and are often employed to make measurements in the ultraviolet and infrared regions too. So presence of any colored chemical agents/dyes/pigments can be thus calorimetrically or spectrophotometrically identified by absorbance spectrophotometer and can be quantitatively estimated frequently using a dilute solution at concentrations smaller than one part of the constituent in several hundred million parts of a selected coloured solution of specific solute. While by UV-VIS Reflectance Spectrophotometer, K/S values (i.e., surface colour strength) of any opaque coloured substance can be determined easily by Kubelka Munk Equation [2, 3] from the measured reflectance values at different wavelength by measuring intensity of reflected diffused light beam or intensity of specular reflected light beam.

Besides surface colour strength, the colour difference and other colour interaction parameters [2, 3, 5, 6] like Total colour differences (DE), Lightness/Darkness (DL*), Red-ness/Green-ness (Da*), Yellowness/Blueness (Db*), Changes in Chroma (DC) and Changes in hue (DH) can be calculated by CIE formulae [1–4]. Also, non-coloured surface appearance properties of any flat sample including textile fabrics can be determined easily in terms of whiteness index, yellowness index, and brightness index values using appropriate and respective formulae of CIE/ASTM or other standards [1–3, 5, 6] to compare any changes in its surface texture for any chemical treatment or physical intervention on the sample, which is very useful for industry.

Limitations and Cares for measuring Reflectance/Surface Colour parameters of solids:

i. **Calibration of the instruments and calibrated dyed samples**: The UV VIS Reflectance spectrophotometer instrument need every day at the start calibration with standard white tiles to combat decayed power of illuminated lamp day by day for correct results. Also, for good colour matching results, the calibration dyeing samples must be prepared with great care at specific dyeing conditions in the laboratory and productions lab unit, which do not differ in various respects viz., checking purity of dye to give same exhaustion, pre-decided M:L ratio, dyeing additive and auxiliaries %, dyeing conditions, dyeing machines settings, exhaustion of dye etc.

ii. **Type of substrate and class of Dyes:** Change of type of fibres or class of dyes changes surface colour strength (K/S Values) data. So, to compare K/S values, the type of fibres, fabric construction, type of dye and dye class should be same. But it is very difficult to maintain the required properties of the substrate from lot to lot as regards the quality of the fibres, yarn structure, fabric construction, colour, heat setting, pre-treatments and dyeing conditions etc. which in turn changes the dye uptake properties of the textiles and give variation in final colour yield/surface colour strength values. It is not at all advisory to prepare the basic calibration data of any particular type of textile fibres with a particular class/type of dyes under all such variations [7, 8]. Changes in dye even within the same class, due to change of groups and conjugation length etc., the dye colour is shifted either bathochromic shift (shift of absorption maxima to longer wavelength i.e., redshift) or hypochromic shift (shift of absorption maxima towards shorter wavelength i.e. blue shift) occurs. Similarly, changes in dye molecules with changes in auxo-chrome type may cause hyperchromic effect (i.e., increase in the intensity of absorption) or a hypochromic effect (decrease in the intensity of absorption) occurs.

iii. **Instrumental colour value vs Human perceived colour**: For colour measurement and matching, formulations generated by the computer-aided colour measuring system are based on minimum colour difference value within given tolerance & least metamerism value in CIELAB scales. But always instrumentally obtained CIELAB tolerances do not confirm exactly with human perceived colour differences and variation persists.

iv. **Failure of Kubelka-Munk Equation in some specific cases:** The measurement of K/s values following kubelka–munk equation does not holds good strictly on extra glossy or fluorescent samples. For too Dull shade or too bright shade sometimes behave differently and do not follow K-M(Kubelka-Munk) theory.

v. **Linearity of Plots between** K/S values vs. concentration of dyes sometimes deviates from linearity and therefore must be checked beforehand. For practical purposes, if non-linear curve appears in the plots between K/S values vs. concentration of dyes, it is needed to re-dye to recheck linearity or to modify the curve by simply eliminating data of one or two erroneous dyed samples to make it linear for precision colour measuring and match prediction. Also, some mordanted textile samples dyed with a variety of natural dyes, do not show linearity for plots between K/S values vs. concentration of dyes.

vi. **Instrumental settings and type of Illuminate sources used**: Test results of instrumental colour values and matching precision varies with changes of instrumental settings for type of CIE standard illuminate used, viewing area (large or small) used, choice of Specular or diffused reflectance measurement, Choice of UV Light on or Off (either UV or VIS or both UV and VIS range of wavelength, Choice of user-chosen wavelength or at deterministic maximum absorbance wavelength, i.e. λmax which are not same for all type of dyes/pigments and hence, there is instrument to instrument metamerism results, which with respect to illuminate type varies too large. Moreover, Instrumental accuracy is to be checked on a daily basis or at least periodically in a week or month.

vii. Some special cares needed during measurement of colour values of solid dyed textiles

a. If the dye uniformity is not up to the acceptable level, the measurement of K/S values will vary a large resulting higher coefficient of variation of K/S data for non-uniform dyeing i.e., Un level dyeing (in general more than 5% coefficient of variation of K/S data is taken as un-level dyeing for all type of dyed textiles).

b. During mounting of solid coloured textile yarns or fabrics, background opaqueness of the sample is to be assured for correct results (So, nos. of folds required in the sample to obtain opaqueness, are to be pre-decided and to be kept constant in all the measurements).

c. During mounting of solid coloured textile yarns or fabrics, changes in the sample orientation (warp wise or weft wise vertical or horizontal measurement or changes of side of the textile fabrics (Colour value in one face of fabric usually differ from other face) differs colour strength values. So, warp or weft wise orientations/and face or backside facing measurements of colour values are to be pre-decided and not to be changed throughout all the colour value measurements of all samples to compare.

d. Some chemical/biochemical treatment before dyeing may alter the surface texture and hence changes scattering value of the sample and hence deviations in K/S value measurement occurs. So, care should be taken to avoid such treatments which changes texture of the sample and alter K/S values to a large extent.

e. Defects in fabric (Any defect of the fabric on surface may cause variation in colour value) like Slabbing/snarls, patchy dyeing, dyeing warp or weft bar etc. causes such variations. So defective fabrics must be avoided during measurements of surface colour values.

3. Colourimetric evaluations for process/product control of dyed textiles

Different types of selective colorimetric evaluation methods are described one by one in brief with examples/case studies with experimental data below mentioning the importance of each method.

3.1 Colorimetric identification and estimation of purity/concentrations of any colorants/dyes/pigments using UV: VIS absorbance spectrophotometer

An optical UV-VIS absorbance spectrophotometer records the absorbance values at different wavelength range at which absorption occurs, together with the degree of absorption at each wavelength and thus a pictorial curve of wavelength (X-axis) vs. Absorbance (Y-axis) called UV-VIS spectrum of that solute from its very dilute solution (preferably 1/100th dilution). The resulting UV VIS spectrum is presented as a graph of absorbance (A) versus wavelength showing maxima (λmax) and minima (λmin) of absorbance at different wavelength in both UV and visible region.

Solute molecules absorb ultraviolet or visible light from a monochromatic beam of incident light beam and the rest are transmitted through the solutions of fixed path length (b or d) in a cuvvete/quartz cell holding the sample solution. As optical density or absorbance is directly proportional to the Path length, *b*, and the concentration, *c*, of the solute/absorbing species of the coloured solution, *Beer's -Lambert Law stands here as*, $\boldsymbol{A = \varepsilon bc}$, (where $\boldsymbol{\varepsilon}$ is a constant of Proportionality, called the *absorbtivity Constant) and is Optical density/Absorbance.*

Different solute molecules absorb UV-VIS light/radiation of different wavelengths depending on its chemical nature and structure with or without interference, if any, as depicted by corresponding absorption spectrum showing absorption peaks and troughs/bands according to the chemical structural groups present in the respective solute molecules present in the coloured dilute solution. Thus, the UV-VIS spectral scan (For absorption or optical density) of a particular-coloured compound/dyes/pigment at a particular wavelength (at λmax) is deterministic and identifiable instrumentally, which is the basis of the identification and estimation of purity/concentrations of any colourants/dyes/pigments by UV VIS spectrophotometric (Absorbance) evaluation.

CASE STUDY 1: As a case study, UV–VIS Absorbance spectrophotometric method of determination of purity and concentrations of rubia/madder as a natural colorant is discussed:

Calibration curve (**Figure 1**) is prepared by using 1,2,3,4,5,6 to maximum of 10 mg of natural *Rubia*/madder (Madder or Manjistha) containing manjishthin and purpurin as natural colourant powder per ml of methanol, after extracted in aqueous solution and purified by Soxhlet extraction under methanol. The solutions were filtered through Whatman filter paper-40 and then used for UV Calibration method and the absorbance was measured at 426 or 430 nm as λmax.

Once the calibration curve is ready in UV VIS absorbance spectrophotometer screen or manual graph paper, the unknown solution of the same compound having unknown quantity of solution is placed in UV VIS scanning taking 10 ml of sample solution of unknown concentration, diluted to known times i.e. say 50 to 100 times until a very fent colour appears in the solution and from that dilute solution 2–3 cc is poured in quartz cell of sample solution and mounted in UV-VIS absorbance Spectro, to measure its Optical density/absorptivity values. Once the absorptivity/ Optical density values of sample of unknown concentration are obtained, the

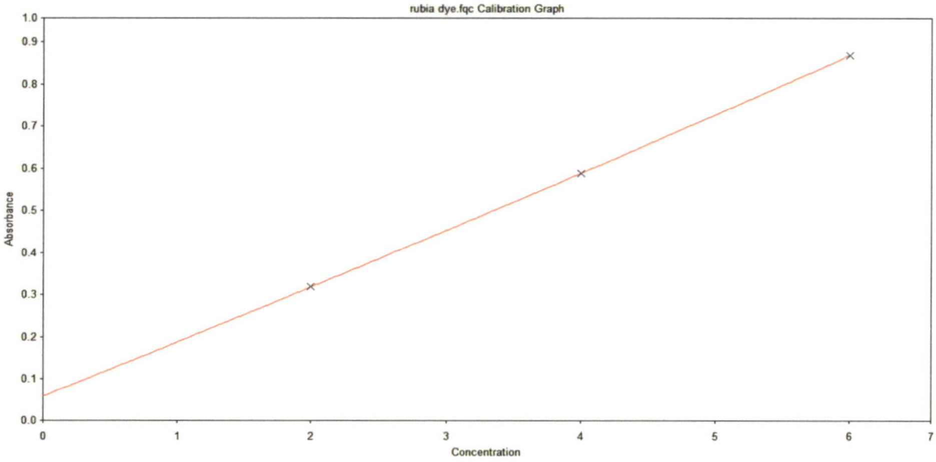

Figure 1.
Calibration curve for the Rubia *dye natural colorant for determining concentration of a solution of unknown concentration of colour component extracted from Rubia/Madder.*

concentration can now be easily obtained by putting measured absorbance or OD (optical density) values in calibration curve of **Figure 1**, to find the Concentration/purity of the content of rubia/madder coloured component in it in specific unit, after correcting the value with dilution factor and converted into proper unit like % or g/lit etc. as per requirement. .

3.2 Identification of colorants/dyes/pigments powder or from its dyed textiles by using UV: VIS absorbance spectrophotometer

CASE STUDY 2: Identification of Natural dye Madder/*Rubia* as a natural colorant is discussed below and is also compared with the synthetic Alizarin coloured compound to distinguish them by UV-VIS Spectrophotometric evaluation method.

The two red dyes– [(i) Natural Rubia (Manjishtha/Madder) which contains manjisthin (similar to alizarin as coloured compounds) in its natural extract and (ii) synthetic alizarin coloured compounds as synthetic same coloured dye] were weighed separately (0.1 gm) and dissolved in 1000 ml dichloromethane/methanol and then wavelength scan under UV-Visible absorbance spectrophotometer was taken for both. For visible spectral analysis, this solution may be used, but for UV spectral analysis this solution needs to be further diluted by 5–10 times for better results. Comparative Identification of Synthetic alizarin and madder (Rubia) as natural colourant/dye, by this UV VIS spectral analysis, involves a comparison of the minute details of UV–VIS peaks/bands of UV VIS spectrum (at λmax) of *Rubia*/Madder as natural colourant and synthetic alizarin red colour. Such a Comparative spectral analysis of both with corresponding UV-VIS peaks is compared in **Figure 2**. The details method of identification of rubia/madder as natural colourant is available in new IS standards **17,085:2019** [9] in Annexure-E as one of the confirmatory tests.

Thus from **Figure 2** Comparative analysis of UV-Vis Spectrum of Natural Rubia/Madder extract and Synthetic alizarin (as shown in **Figure 2** indicate that Natural *Rubia/Madder colourant shows uv–vis peaks at* 250 nm (with 0.954 OD) and at 491 nm (with 0.171 OD), while Synthetic Red alizarin shows *uv–vis peaks at* 250 nm (with 1.38 OD) and 426 nm (with 0.309 OD). This different OD at 250 nm in UV zone and Peaks in Visible Zone at two different wavelengths (at 491 nm with 0.171

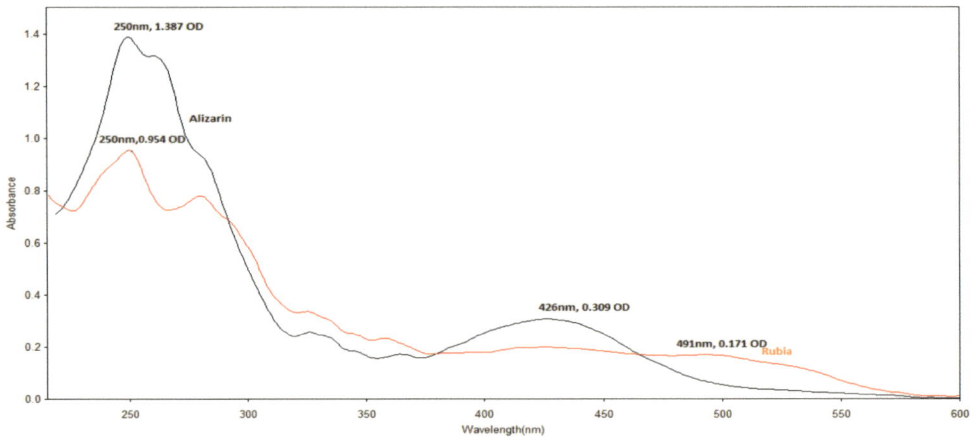

Figure 2.
UV spectrum of Rubia *and Alizarin dyes: [Source -IS standard- 17,085: 2019 [9]].*

Specific Wavelength (nm) and Sample	Peak Reading at Specific wavelength (nm)		Results with inference (describing difference in UV VIS spectra between the two samples taken)
	For Synthetic Red Alizarin	For Natural *Rubia/Madder colorant*	
250 nm and 426 nm for synthetic alizarin	250 nm (1.38 OD) and 426 nm (0.309 OD)	—	The pattern of the peaks in UV and Visible region are very different for the two samples **indicating and confirming their presence/absence in corresponding samples. (natural Rubia/Madder colourant and synthetic alizarin)**
250 nm and at 491 nm for rubia/madder as natural colourant	—	250 nm (0.954 OD) and 491 nm (0.171 OD)	

[Source-IS standard- 17,085: 2019 [9]].

Table 1.
UV VIS spectral peaks analysis of natural Rubia *colorant and synthetic alizarin.*

OD for Rubia/madder natural colourant and at 426 nm with 0.309 OD) are identifying factors confirming presence or absence of them, as shown **Figure 2** and **Table 1**.

Individual UV VIS absorbance Spectrum at visible region only at 390–700 nm, when is partly enlarged for 390–450 nm, it is also observed that the UV–Vis absorbance spectrum of aqueous solution of natural Rubia/Madder extract (extract of Indian Madder i.e., natural manjishthin) also shows small hump like peaks at 398 nm (with 0.801 OD) and also indicating large hump like peak at also 426 nm (with 0.838 OD), which are vivid from **Figure 3**. Therefore, the appearance of the above said two respective peaks in the said wavelength region lead to indicate the presence of natural Rubia/Madder with manjisthin (not synthetic alizarin), which is more clearly understandable in the enlarged peak of highlighted part of UV–VIS Spectrum (**Figure 3**) of extracted solution of *Rubia-cordifolia (used as Natural dye) dyed cotton textiles. Thus, even if Rubia/Madder shows peak at 426–430 nm showing* λmax at around 426 nm (with 0.838 OD) i.e., at the same wavelength where synthetic alizarin has also shown its peak at 426 nm (with 0.309 OD), but OD values are different and thus these two red dyes are easily distinguishable by this method.

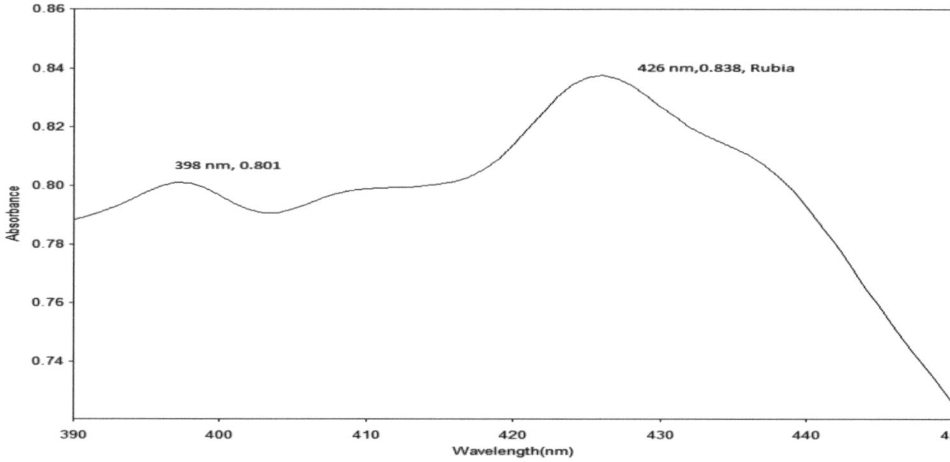

Figure 3.
Part of enlarged UV-Vis Spectrum of Natural Rubia/Madder colorant (containing manjisthin).

3.2.1 Confirmation of identification of natural RUBIA*/madder as compared to synthetic alizarin from UV: VIS spectral scan analysis*

Optical density/absorbance at λmax for extract of natural *Rubia/Madder* colourant and synthetic alizarin dyes are thus found to be quite different from UV VIS Spectral analysis. In the UV–VIS spectrum, although the peaks are at 250 nm, they have shown different optical densities/absorbance values (**Figure 2** and **Table 1**). In the visible region, peaks at 426 nm for alizarin and peaks at 398 nm (0.801) and 426 nm (0.838) for Natural *Rubia*/madder dye-containing manjishthin (a natural coloured compound similar to alizarin, but bio synthesised as natural colorant in Madder/Rubia plant in association with other natural ingredients) are the characteristic peaks of differentiating alizarin and manjisthin (from natural rubia/madder). There are other different methods available for identification of natural Madder/Rubia colourant by either HPLC-DAD analysis or LC–Ms/UPLC-MS analysis or also by FTIR analysis and NMR analysis, some of which are detailed in IS standard**s-17,085:2019** [9] and International ISO Standard: **ISO/Standards-22,195-1-2019** [10] published in recent past in 2019.

3.3 Colour quantification and measurement of surface colour strength (K/S values) of any dyed textiles using UV–VIS-reflectance spectrophotometer

Colour quantification in mathematical term is necessary to develop a systematic understanding of the principles of colour perception and measurement for understanding the differences between colours of two samples i.e., match and mismatch for any method of colour encoding/imaging and communications, to give a more realistic picture for colour reproduction. Hence, TRISTIMULUS VALUES (X, Y, Z) are defined as three coordinates to define any colour for communications, where X, Y and Z values are as follows:

Thus, Tristimulus values X, Y, Z can be calculated from measured total reflectance of a textile or similar flat surface with its derivative formula function as shown below [2–3, 6–8]:

$$X = \sum P_\lambda x_\lambda R_\lambda \tag{1}$$

$$Y = \sum P_\lambda y_\lambda R_\lambda \tag{2}$$

$$Z = \sum P_\lambda z_\lambda R_\lambda \tag{3}$$

where $P_{\lambda=}$Spectral power distribution of standard source, $R_{\lambda\ =}$ Spectral reflectance of substrate and $x_{\lambda.}$ $y_{\lambda.}$ $z_{\lambda=}$colour coordinates/factor of standard observer for red, blue and green.

For ease of working, colours are redefined from TRISTIMULUS values to CIE Chromaticity coordinates (x, y and z instead of Capital X, Y, Z as tristimulus values), which can be plotted in two-dimensional plot. These new CIE chromatically coordinates (x, y, z) can be defined as follows.

$$x = \frac{X}{X+Y+Z} \tag{4}$$

$$y = \frac{Y}{X+Y+Z} \tag{5}$$

$$z = \frac{Z}{X+Y+Z} \tag{6}$$

and

$$x + y + z = 1 \tag{7}$$

From Eq. (22), i.e., x + y + z = 1, the value of anyone CIE chromaticity coordinate can be determined from the values of other two CIE chromaticity coordinates, i.e., the third one can be determined easily from first two.

Still, as Plot of Dye Concentrations Vs Reflectance (R) are non-linear and non-additive, Tristimulus values X, Y, and Z are interdependent on one another, and CIE chromaticity coordinates are still two factors dependent variables to get third one. HUE, value chroma are also 3 coordinates based, it is difficult in practice to control all those multivariate/factors/colour parameters simultaneously to get a precision match of colour.

So, quantification of colour was finally made by Kubelka and Munk [2, 3, 6–8], where K/S value (surface colour strength) is defined as follows:

$$\text{Surface colour strength}\left(\frac{K}{S}\right) = \frac{\text{Co–efficient of absorption}}{\text{Co–efficient of scattering}} = \frac{(1 - R_{\lambda max})^2}{2R_{\lambda max}} = \alpha C_D \tag{8}$$

where K is the coefficient of absorption; S, the coefficient of scattering; and $R_{\lambda max,}$ is the Reflectance value at maximum absorbance wavelength (λmax) and C_D is the dye concentration and α is the constant. Moreover, K/S Vs Dye concentration plots are linear, and K/S is additive in nature.

For Additive nature of K/S value, for use of mixture of colourants/dyes at different concentrations c1, c2 and c3 respectively for dye1, dye2 and dye3 K/S values of resultant fabric may be written as:

$$(K/S)_{MAX} = (K/S)_{subs} + (K/S)_1 + (K/S)_2 + (K/S)_3 + \cdots + \tag{9}$$

$$= (K/S)_{subs} + \lambda_1 C_1 + \lambda_2 C_2 + \lambda_3 C_3 \tag{10}$$

Thus, handling of K/S values become much easy to match colour, as because K/S is treated as a single variable i.e., it operates on a single constant theory (scattering remaining constant for same fabric and dye sample) and K/S is directly proportional to dye concentration in linear and additive relationship.

For dyed textiles/clothes, it is pre-assumed that dyes on specific textile fabric do not add or substract, i.e., change scattering and K is the sum of absorption of dye stuff on dyed textiles and therefore, it is only dye absorption values of dyed textile substrate (if textile substrate remained unaltered/fixed). So, it may be considered that for dyed textiles, K/S directly varies with concentration of dyes linearly and scattering of dyed textile substrate is independent of dye concentration (which is not the case for pigments in paints for wall colours). So, in textile it is single constant theory of colourmatch prediction through K/S values, as most widely applicable colour parameter for colour quantification, measurement and colour matching of textiles. So, for the particular dyed textile sample (with same fibre material, yarn parameter and fabric construction/surface finish remain unaltered) scattering value is assumed to be constant.

Thus, higher is the K/S value, meant higher is the dye absorption in textiles, meant higher absorption value of dye thus signifying or indicating higher dye uptake, but this measurement is surface colour strength, not bulk dye uptake, which can only be determined by extraction of colour from dyed textile samples and then analysis of optical density or absorbance values in absorption spectrophotometric analysis of coloured liquid.

Dye Uniformity in terms of CV % of K/s values at minimum 10 different points may be expressed for deciding factor for level/unlevel dyeing. CV % of K/S values within 5% value is considered as acceptable for level dyeing and more than 5% values (CV % of K/S values) is considered as un-level dyeing leading to rejection of the sample.

3.4 Measurement of colour differences by estimating DE, DL*, Da*, Db*, DC and DH

Colour attributes of a human perception consisting of any combination of chromatic and achromatic content in terms of differences in combination of red, blue and green sensation of human eye (as shown in **Figure 4**) alters change in predominating hue which can be described by chromatic hue names such as yellow, orange, brown, red, pink, green, blue, purple, etc., or by achromatic colour names such as white, grey, black, etc., and is associated with some other attributes like bright, light, dark etc., hence colour differences in between two samples arises by value of these attributes of human perception or instrumental measurements. Measurement of colour differences is important for judging two nearer coloured

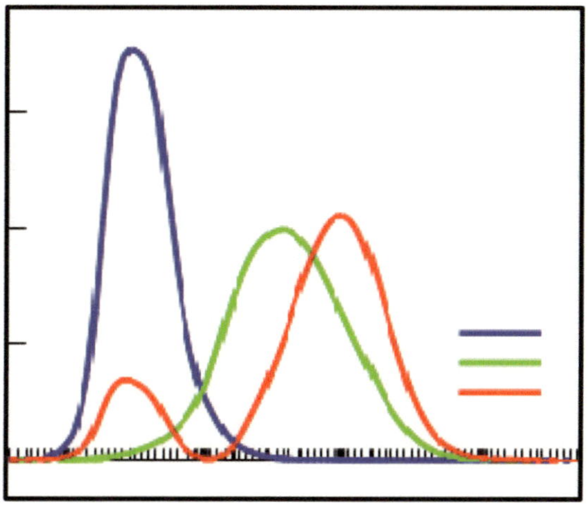

Figure 4.
Red-Blue-Green perception of colour.

samples as match with degree of matching or mismatch. It is Judged by differences in light and dark (ΔL*), Redness or Greenness (Δa*) and Blueness or yellowness (Δb*) as CIELab* colour difference coordinates in CIE colour difference space diagram to determine the total colour difference values (in terms of Δ*E**) from respective CIE-Lab equations following CIELab* standard-1976, which are measurable in UV VIS Reflectance Spectrophotometer and Associated software attached to computer aided colour measurement and matching system.

Thus, according to CIE (Commission International de eclairase, Paris) 1976, total colour difference values (in terms of DE* or Δ*E**) as obtained from individual DL* or ΔL* (Light or dark), Da* or Δa* (Redness or greenness), and Db* or Δb* (Blueness and yellowness) values makes it easy to compare the colour difference values in between two nearer match samples (standards and produced) and this gives a degree of matching according to tolerances set for these attributes of colour as well as gives opportunity to correct shade by adding exactly required colour/dyes to improve less red, less green, less yellow or less blue sample to solve its light and dark by adding white or black as well during dyeing production.

The above said terms DE* or Δ*E* represent total colour difference,* DL* or ΔL* by 0–100 scale representing lightness and darkness, Da* or Δa*, if positive represents redness and if negative represent greenness and Db* or Δb*, if positive represents yellowness and if negative represent blueness by their positive and negative values respectively, as shown in Eqs. 5 to 8 and pictorially is shown in **Figure 5**.

The above said CIE colour differences equations are depicted below for ease of understanding:

$$\Delta E = \left[(\Delta L^*)^2 + (\Delta a^*)^2 + (\Delta Lb^*)^2\right]^{½} \tag{11}$$

where,

$$L^* = 116(Y/Y_a)^{⅓} - 16 \tag{12}$$

$$\Delta L^* = L_1^* - L_2^* \tag{13}$$

$$a^* = 500\left[(X/X_a)^{⅓} - (Y/Y_a)^{⅓}\right. \tag{14}$$

$$\Delta a^* = a_1^* - a_2^* \tag{15}$$

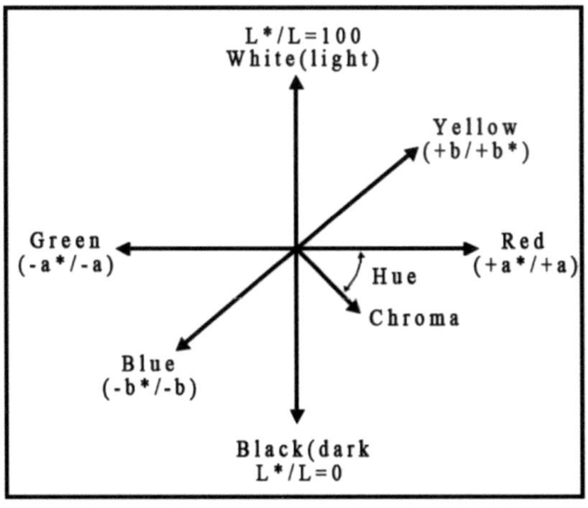

Figure 5.
*CIE L*a*b* colour difference space diagram. By human eye (wavelength vs. intensity).*

$$b^* = 200\left[(Y/Y_a)^{1/3} - (Z/Z_a)^{1/3}\right] \tag{16}$$

$$\Delta b^* = b_1^* - b_2^* \tag{17}$$

Chroma, (psychometric chroma) values in CIELAB colour space can be calculated as follows:

$$C^*_{(ab)} = \left(a^{*2} + b^{*2}\right)^{1/2} \tag{18}$$

$$\Delta C^* = C^*_{1(ab)} - C^*_{2(ab)} \tag{19}$$

where, $C^*_{1(ab)}$ and $C^*_{2(ab)}$ are the chroma values for standard and produced sample.

CIE 1976 metric Hue-Difference (ΔH) for CIELAB system can be calculated as follows:

$$\Delta H_{ab} = \left[(\Delta E^*_{ab})^2 - (\Delta L^*)^2 - (C^*_{ab})^2\right]^{1/2} \tag{20}$$

Moreover, Brightness is another additional colour attribute associated with perception of colour differences. This attribute of visual sensation of colour gives an additional visual perception that appears to be more or less intense or luminescence i.e., this visual stimulus appears to emit more or less light from specific hue of colour and differs from one another.

Brightness Index (BI) as per ISO-2469/2470–1977 method [11] can be calculated by following ISO formula for this:

$$\text{Brightness Index} = \frac{\text{Reflectance Value of the Sample at 457 nm}}{\text{Reflectance Value of the Standard white diffuser(white tiles)at 457 nm}} \times 100 \tag{21}$$

Application of fluorescent brightening agents to white textiles show an additional higher reflectance value more than 100 and up to 150. Though the sample appears to be still whiter as usual, there is emitting of more reflectance of incident light in the bluer zone and the appearance thus changes its chroma towards blue increasing its more whiteness and brightness, where brightness value may be represented or expressed in quantitative term by ISO standard method. Conversely, yellowing of white textiles by chemical treatment or by heat scorching or by any type of degradation by exposure to light or by gas fading etc. can blur the brightness value of the white or dyed sample. Thus, along with colour differences like DE, DL, Da, Db, this Brightness index (BI) as another additional colour attributes related to surface appearance properties of textiles have immense important role and simply high or low BI values an important colour surface appearance parameter too in defining the colour quality of any textile fabric.

A recent newer concept of defining colour differences by Colour Difference Index (CDI) values as a measure of dispersion of colour values at different points from all angle of instrumental measured variation, depending on dyeing process variables, to understand the combined effects of different dyeing process variables by a single parameter, is defined [12] taking only the magnitudes of the respective ΔE, ΔC, ΔH and MI values (irrespective of their sign and direction), to calculate CDI values using the following empirical formula (Eq. 22).

$$\text{Colour Difference Index (CDI)} = \frac{\Delta E \text{ X} \Delta H}{\Delta C \text{ X MI}} \tag{22}$$

Mordant Concn. (%)	*K/S* At λ_{max}	ΔL	Δa	Δb	ΔC	ΔH	ΔE	MI (LABD)	CDI
5	2.24	−12.52	3.77	8.25	8.95	−1.49	23.35	1.19	3.26
10	2.65	−15.61	4.29	8.24	9.07	−2.00	22.87	1.34	3.75
15	3.89	−20.90	4.73	7.53	8.56	−2.42	17.97	2.11	2.41
20	3.15	−16.51	4.84	12.09	12.92	−1.64	18.06	1.69	1.35
25	2.45	−12.90	4.53	10.81	11.60	−1.68	20.57	1.54	1.94

Table 2.
Effect of Mordant concentration on Colour Strength and Colour Differences for dyeing silk fabric with tesu (containing butein) extract as natural colourant.

Higher the differences in between maximum and minimum CDI values, higher is the dispersion of colour values at different points i.e., colour values are more widely dispersed, and that variable become critical for reproducibility for such dyeing. So, lower the differences in between maximum and minimum CDI value in one set of dyeing for particular dyeing process variables or use of mixture of same set of binary mixture of dyes, better is the match with lower dye dispersion in such cases of colour match CDI value below 5 is acceptable and good and below 1.0 is considered as excellent.

CASE STUDY 3:

The above shown data in **Table 2** on colour parameters, obtained in a study on use of different mordant concentration yields different surface colour strength(K/S) showing reasonable differences of Colour values in terms of ΔE, ΔL, Δa, Δb, ΔC, ΔH, MI (LABD) and CDI values indicating the inter-dependence of colour strength and other colour interaction parameters of tesu dyed silk fabric, clearly showing the role of increasing mordant concentrations up to 15% for higher K/S values, with maximum ΔE, and medium CDI, while increase of Mordant concentration beyond 15–25%, gradually reduces colour strength but increases colour dispersion with lowering of CDI.

3.5 Computer aided colour match prediction of textiles and others by using UV: VIS reflectance spectrophotometer and colour measuring/matching software for producing any standard shades

Colour matching of two samples are considered as fully satisfactory, if any one of the following 3 conditions are achieved with plus-minus mutually accepted tolerances values of their colour differences in CIELab attributes as follows:

Thus, to become colour of produced sample = colour of given standard sample, following should be the conditions be satisfied - i.e., below given conditions (1)–(3).

1. (X_{SL}, Y_{SL}, Z_{SL}) values of produced sample = (X_{SD}, Y_{SD}, Z_{SD}) values of given standard sample where X, Y & Z are the tristimulus value of Sample (SL) and Standard (SD)

2. $(\text{Reflectance})_{SL}$ value at 400 to 700 nm of produced sample = $(\text{Reflectance})_{SD}$ value at 400 to 700 nm of given standard sample

3. $(K/S)_{SL}$ value of produced Sample = $(K/S)_{SD}$ value of given standard sample, where $K/S = \alpha C$.

3rd Conditions are easy to check and achieve, as it is additive in nature and Dye Concentration vs. K/s values plot are linear and is predictable from sample database by computerised algorithm.

For computer aided colour matching theory [2, 6–8], for a shade from mixture of multiple colourants (say 3 colourants), following three equations are to be solved as a function of dye concentrations of the colourants (1, 2,3 or n) and to be checked by measuring tristimulus values or reflectance values or K/s values with measurement of DE*, DL*, Da* and Db* values under different standard illuminants.

$$f(c_1, c_2, c_3) = x$$
$$f(c_1, c_2, c_3) = y$$
$$f(c_1, c_2, c_3) = z$$

where x, y, z tristimulus values of standard given sample are to be matched with the matched dyed textile sample to be produced, by using say 3 different dyes with respective concentrations of those 3 selective dyes indicated by c_1, c_2 and c_3. For determining/predicting these selective concentrations of specific dyes to get a specific match of colour, In practice, the reflectance values of standard sample at 400 to 700 nm are initially measured from standard dyed textile substrate and those reflectance data are processed through computer aided software to generate matched K/S Values, within tolerance set for specific L, a and b colour difference parameters and DE total colour difference parameter to match for the predicted/produced sample. As K/S values vs. concentration of dyes is linear & additive, so this is used as basic data for handling colour match prediction by computer aided colour measuring cum matching instrument from different companies with application software in built in the system.

Colour matching is always associated with Some practicable values of DE*, DL*, Da* and Db* values, within acceptable tolerances, but is also associated another factor/term called metamerism index (MI), due to measurement of colour values under different conditions of measuring colour values i.e. within varying illuminates or varying observers or varying instruments etc. [2, 6–8].

Thus, only colour difference values do not represent true differences of perceived colour in human eye due to observer's metamerism or even instrumental metamerism or illuminate metamerism etc. An ideal or perfect colour match is called isomeric match i.e., which are always match under all illuminates or under all observers or under all instruments in all the ranges of wavelength values in visible region and then that ideal match is called true isomeric match. While Most of the given standard of colour and produced samples are not at all show isomeric match, there is always some differences in their colour difference results at different wavelength range or otherwise i.e. when two coloured sample (standard and produced sample for colour matching) show match under one illuminant/one observer or one instrument but do not match under any other illuminant/other observer or other instrument at different wave length values is termed as a metameric match. So, it is a challenge to produce a Least metameric match instead of ideal isomeric match. A general metamerism index (MI) value can be calculated using Eq. 23, as follows:

$$\text{General metamerism Index} = \frac{\sum(\Delta R\bar{x})^2}{X^2} + \frac{\sum(\Delta R\bar{y})^2}{Y^2} + \frac{\sum(\Delta R\bar{z})^2}{Z^2}, \tag{23}$$

where ΔR = Difference in reflectance between pair of metamer samples; $\bar{x}, \bar{y}, \bar{z}$ = CIE standard observer colour function X, Y, Z = CIE tristimulus value normally taken for illuminate C. It is average value of colour differences of two specimens under two different measuring conditions.

The Metamerism-Index (MI) indicate the probability of any two near match or matched two samples when show the different colour difference values under changed conditions of measurements like if measured under two different illuminants (represented by the first and second illuminant) or under two different make reflectance spectrophotometer instruments or under any other two different conditions of measuring colour parameters of the said two specific samples by calculating. CIE LAB i.e., LABD metamerism index [2, 6–8], which is represented below in Eq. 24:

$$\mathrm{MI}_{(\mathrm{LABD})} = \left[(\Delta L_1^* - \Delta L_2^*)^2 + \left(\Delta a_1^* - (\Delta a_2^*)^2 + \left(\Delta b_1^* - (\Delta b_2^*)^2\right]^{1/2} \quad (24)$$

ΔL_1^*, Δa_1^*, and $\Delta b_1^*$are the Delta CIELab* colour coordinates between standard and sample for the first illuminate and ΔL_2^*, Δa_2^*, and Δb_2^* are the Delta CIELab* colour coordinates between standard and sample for the second illuminate interpretation:

If MI is low, the colour difference between the sample pair is the closer and more similar for different conditions of measurement, even under different illuminates or observers or instruments. So, matching of two-coloured samples produced at comparable conditions are to always to minimize to obtain least metameric match for control of colour by using computer aided colour measuring and matching system [7, 13].

CASE STUDY 4: Computer aided colour match prediction for dyeing of textiles: as an Example

Practical Guideline for Colour Match prediction: it is necessary to prepare Company wise Dye Class type and Sample type (Substrate fibre type) database by calibration dyeing [7, 14, 15] of 0.25, 0.50, 0.75, 1.00, 1,25, 1.50, 1.75. and 2, 2.5, 3 percent dyed sample of specific fabric (based on type of fibre) i.e., say- bleached cotton fabric and their reflectance or X, Y and Z Data are to be measured and to be saved as library of database for use for formulation prediction of dye weight % required for colour matching from time to time for given standard sample.

Colour matching tolerances against Standard daylight D65 illuminate, Artificial Tube light -TL84 (A) and fluorescent light (F) are to be set as maximum 1.00 for each light or to be mutually fixed between buyers and sellers in order agreement. If dye cost from lot to lot regular purchase is updated in this system, cost of dyes for different formulations are also calculated and available at fingertips, other dyeing process and utility cost remaining same. Not only it helps to reduce dye inventory and it saves matching time for lab to production trial time with reasonable known combination of dyes and cost involved along with average predicted dE*, dL*, da*, db* values to know the degree of precision of colour matching, below is the example of one colour match prediction formulation using computer aided colour matching system with database for different class of textile dyes already fed in (the present example is colour matching formulation of cotton fabric with reactive dyes database, as given in **Table 3**.

Thus, the above predicted 2 formulations indicate that formaulation#1 is less metameric as understood from comparison of their dE*, dL*, da*, db* values, and cost wise Formulation#2 is least cost match,

3.6 Estimation of compatibility between two colorants to use for compound binary shades

Compatibility between any two same class of dyes can be judged by different methods, such as (i) comparative subjective visual assessment of the degree of on-

Standard Id = Coloured Cotton Fabric-C 12										
RFl	3.65	3.90	4.46	5.87	7.44	8.77	9.32	10.61	11.56	12.71
DATA For Std.	12.33	11.14	10.33	9.34	8.10	7.82	6.98	6.53	5.39	4.34
Dye class used	= Reactive Dye database for white cotton									
Dye ID# used	1,3,4, 6, 7,9,10, 11and 12 from data base									
Substrate ID#	3, Enzyme pre-treated Bleached Cotton									
TOLERENCES	dE* for D65 Light = 1.00							dE* for Artificial Light = 1.00		
	dE* for Fluorescent Light = 1.00									
ID#	Colorant	Amount	Per cent			da*	db*	dL*	dE*	Rs
Matching Formulation Generated by computer Aided Color Matching System										
Formula#1										
3	R Red M3B	0.15	0.15		D	−0.0	−0.0	0.0	0.0	56.51
6	R Brown 5R	0.68	0.68		A	−0.42	0.2	0.10	0.23	227.73
10	R Procian Blue 2R	1.41	1.41		F	−0.62	0.4	0.33	0.66	495.00
		2.14	2.14							778.24
Formula # 2										
3	R Red M 6B	0.13	0.13		D	−0.0	−0.0	0.1	0.01′	48.53
11	R Grey 2R	0.54	0.54		A	−0.52	0.3	0.0	0.36	188.26
10	R Procian Blue 2R	1.28	1.28		F	−0.71	0.5	0.11	0.78	449.36
		1.95	1.95							686.15

Table 3.
Example of a colour match predicted from the database of direct dye for cotton.

tone build up by carrying out a series of dyeing for both dyes to same substrate and checking gradual colour build up by visual assessment, (ii) theoretical prediction of compatibility [16] by comparison of rates of dye by rate of diffusion of dyes by determining diffusion coefficients or by determining time of half dyeing for each individual dye at comparable dyeing conditions (iii) by quantitative assessment of change in hue angle(ΔH) for increasing dyeing time and temperature or increasing dye concentrations [16] under two sets of dyeing for colour built up on specific textile substrate (iv) by comparing the nature of plots of ΔC vs. ΔL or K/S vs. ΔL values for two sets of progressive built up shades as said in point -no (iii) obtained by dyeing with varying dye concentration and also with varing dyeing time and temperature as said in point no-3 using 50:50 of two dyes [17] and (v) quantitative compatibility rating for the mixtures of more than two dyes by colorimetric analysis of actual colour strength developed (not on the basis of dye absorbed) for mixture dyeing in different proportions following Relative compatibility rating (RCR) method [12] by calculating differences of CDI (Colour difference Index) values [17, 18] as a newer empirical index of overall colour differences for dyeing different proportions of two dyes of different pairs of synthetic or natural dyes applied on any textiles.

CASE STUDY 5: Comparison of compatibility of two dyes by comparing the nature of plots of ΔC vs. ΔL or K/S vs. ΔL values for two sets of progressive built up shades by dyeing with variation of dye concentrations (SET-1) and dyeing with variation of Time and temperature (SET-2) using 50:50 of two dyes as well as also Determining compatibility of 2 dyes by Relative compatibility rating (RCR) method by calculating differences of CDI values for dyeing different proportions of any two dyes.

Dyes Selected are: Direct dyestuffs (make: Atul Ltd. (Tuladir)) of four different colours, i.e., Direct Turquoise blue (CI Direct Blue 199), Direct Red (CI Direct Red 31), Direct Yellow (CI Direct yellow 44), Direct Green (CI Direct Green 513).

Dyeing carried out for Conventional methods of determining compatibility, for obtaining plots of ΔC vs. ΔL or K/S vs. ΔL values for two sets of progressive built up shades following selected binary pairs (50:50) of synthetic direct dyes were applied on the 6% H_2O_2 (50%) bleached Jute fine hessian fabric using three pair of following combination of binary pair of direct dyes such as M-11 -Direct Red + Direct Green, b) M-12-Direct Red + Direct Yellow and c) M-13-Direct Red + Direct T. Blue taken in 50:50 ratio in two sets.

In Set I, the progressive depth of colour was gradually built up by varying dyeing time and temperature profile for each pair of dyes (M11, M-12 and M13), three jute fabric samples were dyed laboratory beaker dyeing machine with temperature controller for 10–60 min varying dyeing time period. The said dyed fabric samples were one by one taken out from the respective dye bath at equal interval of 10 min from dyeing temperature of 60°C onwards up to 100°C, maintaining the constant heating rate of 2–5°C/min. The final and ultimate dyed sample was taken out from dye bath after 60 min dyeing time at 100°C dyeing temperature.

In Set II, the progressive depth of shade was obtained by varying total concentration of dye mixture in 50:50 ratio but varying percent application from 20–100% of 1% shade for each pair of dyes, for 3 separate samples of jute fabrics, which were dyed at the at the increments of 20% points of dye concentration at pre-fixed dyeing conditions at 100°C for 60 min. Taking two dyes in equal proportions (50:50).

The colour difference values in terms of ΔE* and ΔL*, Δa*, Δb* and ΔC* for all the above said dyed fabrics using Set I and Set II conditions, against undyed fabric sample as standard for reference, were obtained by individually separate measurement of the colour difference parameters Using UV–VIS reflectance spectrophotometer within built software and computer attached. The compatibility of a selected pair of dyes was judged [16–19] from the degree of closeness and overlapping of two curves ΔC vs. ΔL or K/S vs. ΔL observed using the two sets of dyeing (Set I and Set II) as shown in **Figure 6**.

For Relative Compatibility Rating Newer method of Determining Compatibility of two dyes, 6% H_2O_2 (50%) bleached jute fabric samples were dyed with four direct dyes taken from Atul direct dye of either single or selected binary pairs of direct dyes in varying proportions (100:0, 75:25,50:50,25:75 and 0:100) under specific fixed and comparable dyeing conditions. The results are shown in **Table 4**.

Thus, both of these methods show a similar results, while the method −2 of RCR compatibility rating method is easier and less time consuming and hence has advantages over plotting of K/S Vs DL.

3.7 Optimization of dyeing process variables for dyeing textiles with any synthetic or natural dyes

Dyeing of any textiles, say cotton or jute or any other fibres to be dyed with specific class of synthetic dyes like reactive dye (or even for any natural dyes) need

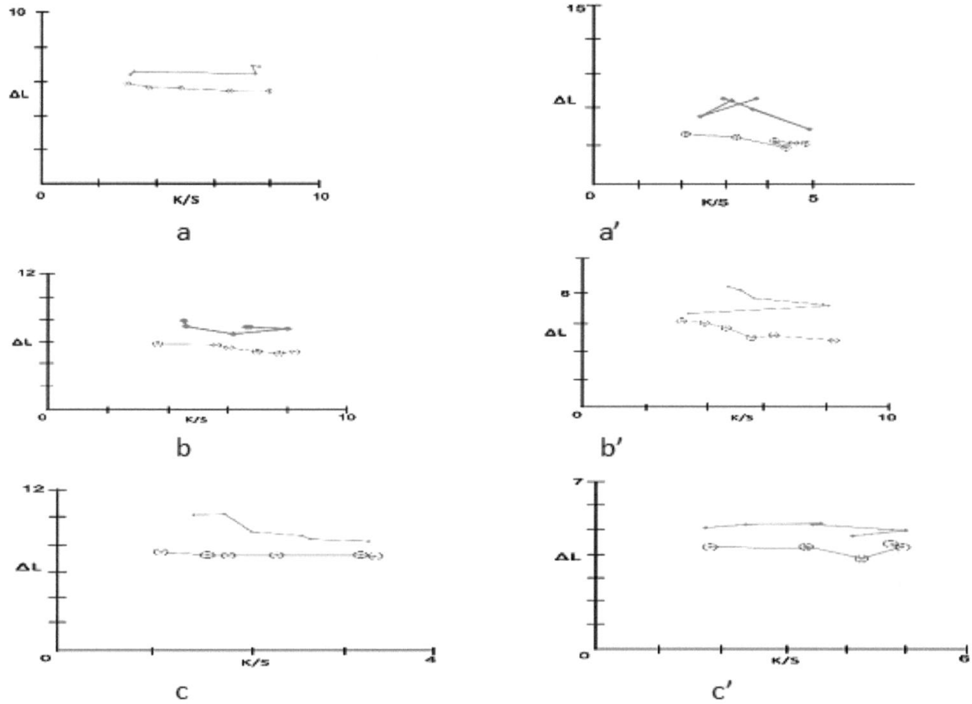

Figure 6.
*Plots showing K/S Vs ΔL curves of (a) M11-D Red: D Green (b)M-12 -D Red: D yellow and (c)M-!3 -D Red: D T. Blue for two sets of each showing M-12 -D Red: D yellow combination has good compatibility, while M11-D Red: D Green combination ahs not so good compatibility or has fair compatibility and M-!3 -D Red: D T. Blue has more or less average compatibility at higher time (**Table 4**).*

to be optimised [14, 15] to derive standard dyeing conditions to obtain maximum surface colour strength (K/S values).

So, it need to have experiments on varying dyeing time, temperature, dye concentration, salt concentrations, MLR and pH etc., so that reproduced and uniform dyeing can be achieved easily.

However, for reactive dyes, Dyeing time has two type -Dye exhaustion time and Dyeing fixation time and similarly dyeing temperature has two dimensions, i.e., Dye exhaustion temperature and Dye Fixation temperature and also for last stage of alkali fixation of reactive dye, addition of soda ash is to be considered, also, besides addition of salt for exhaustion as evident from earlier references [20].

UV VIS reflectance spectrophotometer thus helps by colorimetric analysis of Surface colour strength and other colour parameters, for dyeing of any fibre with specific class of dye by varying conditions of dyeing.

|CASE STUDY 6: Optimization of dyeing process variables for jute dyeing with reactive dyes.

Fabric used: 3% H_2O_2 bleached fine hessian jute fabric having 215 tex jute yarns as warp and 285 tex jute yarns as weft, 64 ends/dm and 58 picks/dm, fabric area density 320 g/m^2 and fabric thickness 0.70 mm, obtained from M/s Gloster Jute Mills Ltd., Bauria, Howrah, was used.

Dyes Selected: (i) Hot brand Reactive Green HE4BD (CI Reactive Green 19), (ii) Hot brand Reactive Orange CN (C.I. Reactive Orange 84) and (iii) Cold brand Magenta (C.I. Reactive Red 11) were used.

Measurement of Colour Parameter: K/S values of differently dyed jute fabrics under varying conditions of dyeing were determined by using computer-aided UV VIS Reflectance spectrophotometer [Premier Colour Scan Instrument Ltd. Mumbai

Dye combination	CDI			Difference in CDI max & CDI min	RCR*	Compatibility grade
	75:25[a]	50:50[a]	25:75[a]			
M11(D Red: D Green)	0.131	0.589	0.039	0.550	2–3	Fair
M12(D Red: D Yellow)	0.149	0.089	0.036	0.113	4	Good
M13(D Red: D T. Blue)	0.037	0.408	0.056	0.371	3	Average

**Based on chart values of differences in max and Min. CDI values [12, 19].*
[a]Proportion of dyes.

Table 4.
Colour Difference Index (CDI) and Relative Compatibility Rating (RCR) for application of selected binary pairs of synthetic dyes of jute fabric.

Make Model SC 5100A] along with associated Colour-Lab plus software employing Kubelka Munk [2, 6–8] equation and CIE-Lab equations against a particular undyed (bleached) sample set as standard followed by calculating the K/S values with the help of relevant software.

The relevant color parameters measured for each sample of varying dyeing conditions are detailed in **Table 5** and plots of each dyeing process variables vs. K/S values are shown in **Figure** 7 for 3 selected reactive dyes applied on jute under varying conditions of dyeing, to optimize dyeing conditions of each dye.

Finally, data in **Table 6** indicate the relevant optimised dyeing parameters for each reactive dyes studied and reported here as optimised dyeing conditions for those respective dyes applied on Jute fabric by conventional reactive dyeing method.

3.8 Precession grading of colour fastness of dyed textiles by colorimetric measurement of total colour difference (dE*) values after fading/staining in colour fastness test procedure

In color fastness test for washing, rubbing or crocking or perspiration, or gas fading or any other agencies, the assessment is done two ways—(i) assessing change of colour/loss of depth of shade and (ii) assessing staining on a same or multifiber white fabric after colour fastness test s by fading under different agencies/conditions as per standard test method and followed by assessing colour loss or staining amount by comparing with two types of grey scale as said. But this assessment is sometimes misleading to one grade upper or lower and is debatable unless quantitative measurement of amount of colour change or amount of staining occur is done and checked not fully depending on visual assessment with the said two types of grey scale.

Colour changing grey scale card consists of colour fastness rating for the colour change with a corresponding decreasing scale of grey chroma, which is standardised in 5-grade levels or nine grades system including half grades, where grade 5 representing the best Colour Fastness and grade 1 representing the worst colour fastness. The middle levels are assessed as half grade: like grade 4–5 and grade 3–4 and then it consists of nine levels.

Similarly, stained grey scale card consists of standard scale of white with a corresponding group of increasing grey chroma having standardised mainly by five grades (1–5), or nine grades system including half grades, where grade 5 implies

Name of variable parameter of dyeing	Parameter varied unit	K/S (Orange CN)	K/S(Green HE4BD)	K/S(Magenta Cold)
Dye concentration (%)	1	7.44	11.9	2.43
	2	7.69	14.98	3.6
	3	9.57	15.31	5.05
	4	**9.87**	**15.70**	**7.51**
	5	9.13	15.53	7.12
Salt (g/L)	30	9.25	8.35	3.08
	40	9.68	11.37	**4.26**
	50	9.73	11.97	3.55
	60	**9.82**	**10.52**	3.89
	70	7.68	11.34	3.67
	80	7.89	12.17	3.97
Dye exhaustion time (Min.)	30	6.69	5.76	2.46
	40	7.07	5.46	2.69
	50	9.46	7.00	3.10
	60	**6.33**	7.43	**3.49**
	70	8.00	**8.44**	2.99
	80	7.79	9.44	3.17
Dye exhaustion temp (°C)	60	6.20	7.35	—
	70	7.65	**8.37**	—
	80	**8.77**	7.03	—
	90	6.81	6.35	—
	100	5.38	5.97	—
Soda Ash(gpl)	10	7.04	2.53	3.34
	12	7.27	3.66	**4.65**
	15	7.82	4.09	4.42
	18	**8.43**	5.69	3.94
	20	8.16	**5.84**	3.66
pH	8	6.39	4.41	3.05
	9	6.42	4.96	3.22
	10	6.78	5.12	3.29
	11	6.95	5.75	**3.38**
	12	**7.25**	**6.01**	2.87
MLR	1:10	10.78	6.30	3.42
	1:20	**11.71**	**8.35**	**4.05**
	1:30	10.63	4.17	3.69
	1:40	9.68	3.38	3.09
	1:50	9.35	3.24	2.41

Plots of dyeing process variables Vs K/S values for three reactive dyed jute fabric dyed with varying dyeing conditions as per ***Table 5.***

Table 5.
Surface colour strength (K/S) data showing the effects of dyeing process variables on colour yield of different reactive dyed jute fabric.

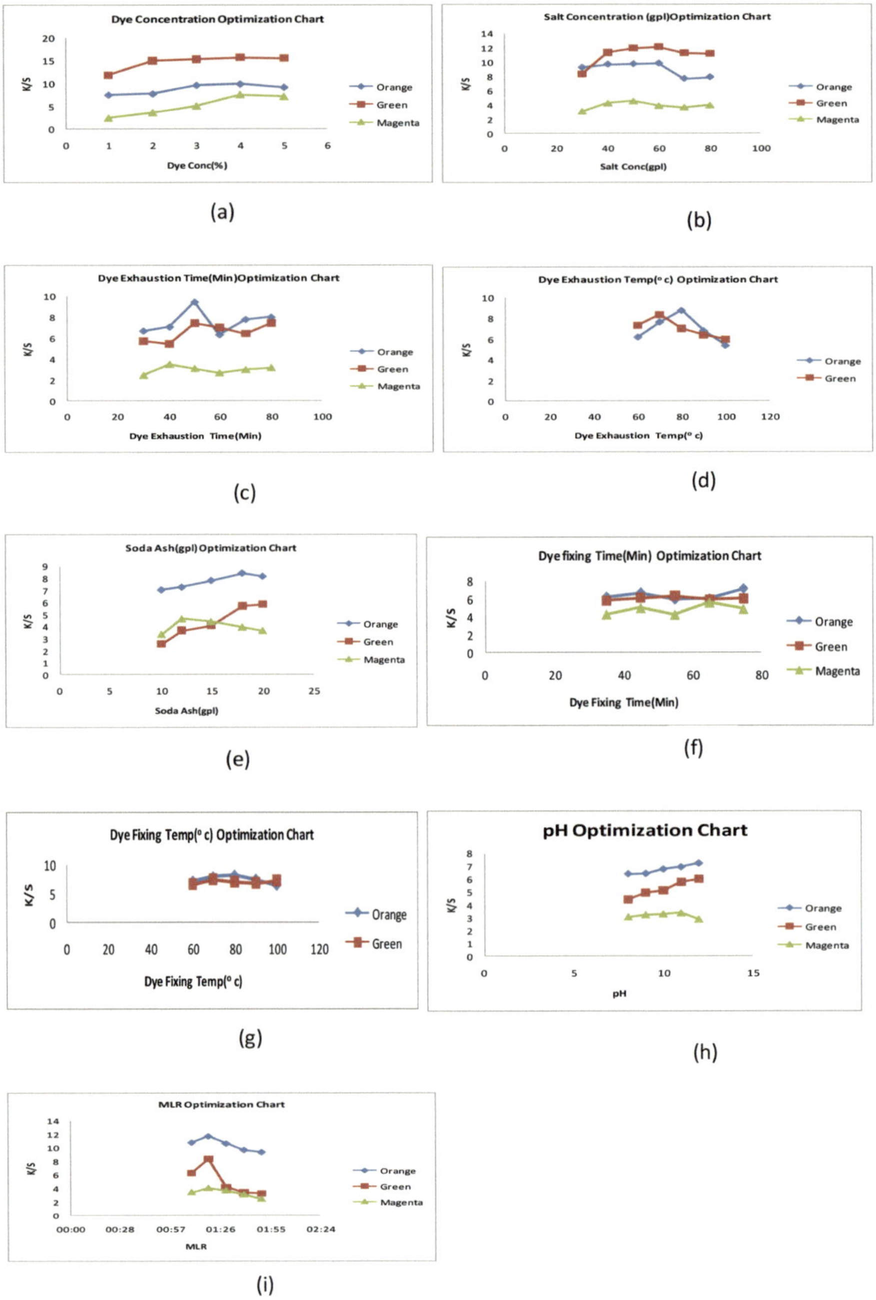

Figure 7.
Plots (a-i) showing dyeing process variables vs. K/S curves for three reactive dyes-for varying. (a) Dye concentration; (b) Salt concentration; (c) Dye Exhaustion Time (Min); (d) Dye Exhaustion Temp (°c); (e) Soda Ash (gpl); (f) Dye fixing time (Min); (g) Dye fixing Temp (°c); (h) pH; and (i)MLR.

virtually no staining representing best colour fastness while grade 1 signifies the worst colour fastness, and the middle grade are assessed as half grade, like grade 4–5 and grade 3–4. But these grey scale grading is comparative visual assessment of grades and may not always be true.

Hence later, as per ISO-105-A02—1993 Textiles- Test for Color fastness test - part -A02, Grey scale for assessing change in color and ISO-105-A03–2019 -

Name of Dye Process Parameters	Re Orange CN	Re Green HE4BD	Re Magenta Cold
Dye Concentration (%)	4	4	4
Salt Conc (gpl)	60	60	40
Dye Exhaustion Time(Min)	50	50	40
Dye Exhaustion Temp(^{0}C)	80	70	-
Soda Ash(gpl)	18	18	12
Dye Fixing Time(Min)	45	55	45
Dye Fixing Temp(^{0}C)	80	70	-
pH	12	12	11
MLR	1:20	1:20	1:20

Table 6.
Optimised conditions of dyeing process variables by conventional method of reactive dyeing of jute fabric using three selected reactive dyes.

Textiles- Test for Color fastness test- part-A03, Grey scale for assessing staining, the quantitative data for dE* values for both types of grey scale are shown in **Table 7** with given tolerances. So precision and correct color fastness grading is now possible matching with the values of measured DE* values after fading/ staining on each type of colour fastness tests under different agencies instead of using visual comparative assessment by grey scales only. Thus, colorimetric measurement of these cases is found to be useful for correct/precision color fastness grading.

3.9 Determination of rate of dyeing, dyeing isotherm and dyeing kinetics parameters by colorimetric analysis

Rate of dyeing can be understood by colorimetric analysis of dye in fibre (rest are dye in solution) at specific dyeing time and its temperature dependence and dyeing isotherm is understood by Din Fibre vs. Dye in solution plots and dye in fibre with respect to different dyeing temperature indicates its bearing on heat of dyeing. All these can be easily calculated by colorimetric analysis of dye absorbed in fibre (out of total dye added in bath) by analysis of dye concentration left in dyeing bath at any time span and even after different dyeing time and temperature, if dye % added in bath solution before dyeing is known. This must be done in UV VIS absorbance spectrophotometer after obtaining calibrated dye concentrations curve for specific dye. Discussion of a case study will bring more clarity in it to understand it practically. Hence, an example of determining rate of dyeing, dyeing isotherm and dyeing kinetics are briefly mentioned as a case study facilitating both offline and on line colour control in relation to computer aided colour control and matching [21] for textiles.

CASE STUDY 9: An example of determining rate of dyeing, dyeing isotherm [Dye in fibre vs. Dye in Solution curves] and dyeing kinetics (determining half dyeing time, heat of dyeing or dyeing enthalpy, bond energy etc] are briefly mentioned here as case study. Relevant data and the rate of dyeing curve [D_f (amount of Dye exhausted to the fibre) vs. t_d (time of dyeing)] for jute fabric for dyeing with madder (also known as Manjistha/Rubia) after double pre-mordanting with 20% harda (myrobolan) and 20% $Al_2(SO_4)_3$ applied in sequence followed by subsequent

Grey scale for assessing change in color		
Fastness Grade	**CIELAB difference**	**Tolerance**
5	0	0.2
4.5	0.8	±0.2
4	1.7	±0.3
3–4	2.5	±0.35
3	3.4	±0.4
2–3	4.8	±0.5
2	6.8	±0.6
1–2	9.6	±0.7
1	13.6	±1.0
Grey scale for assessing staining		
Fastness grade	CIEIAB difference	Tolerance
5	0	0.2
4–5	2.2	±0.3
4	4.3	±0.3
3–4	6.0	±0.4
3	8.5	±0.5
2–3	12.0	±0.7
2	16.9	±1.0
1–2	24.0	±1.5
1	34.1	±2.0

Table 7.
Colour fastness grading in terms of colour difference values (dE) as equivalent to grades of grey scale with tolerances for precision grading of colour fastness assessment.*

dyeing with madder/Manjishtha under a pre-optimized conditions of dyeing are shown in **Table 8** and **Figure 8**.

Relevant Data in **Table 8** also shows the dye exhaustion to the fibre for different dyeing temperature indicating rate of dyeing for application of madder extract on the said double pre-mordanted jute at lower temperature (at 50°C) and at higher temperature (at 90°C), where differences of dye up take at these two temperature are found to be higher at lower dyeing temperature of dyeing and gradually the differences reduces for use of higher temperature, viz. data in **Table 8**.

Relevant curves in **Figure 8**, using data of **Table 8**, indicate that with increase in dyeing time, the dye uptake (D_f) increases measurably up to 60 min of dyeing time and then gradually slows down and almost levels off in between 90 and 120 min. Since, purpurin and manjistin are present as the two main colouring components of in Indian Madder [a natural dye], both these colouring components [having -OH and -COOH functional groups] gradually starts reacting by attachment to mordant with increasing of dyeing time and temperature, while its exhaustion to the mordanted fibre might have levelled off after possible saturation of such dye-mordant-fibre complex forming reaction and possible hydrogen bonding etc. for dye fixation is completed and no further increase in temperature or time can increase dye up take further.

Time (min)	[D] $_f$, g/kg at 50°C	[D] $_f$, g/kg at 90°C
15	1.5	2.6
30	2.7	4.0
45	3.8	5.2
60	4.8	6.1
75	5.5	6.5
90	5.9	6.6
120	6.1	6.6

Table 8.
Dye exhaustion to the fibre [Df] for different dyeing time indicating rate of dyeing for application of Madder/manjistha as natural dye on double pre-mordanted bleached jute fabric.

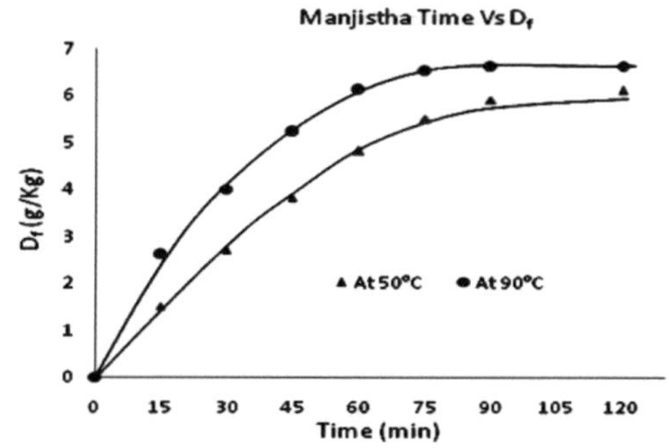

Figure 8.
Rate of dyeing plot as function of time for dyeing of pre-mordanted jute fabric with Madder at 50 and 90°C.

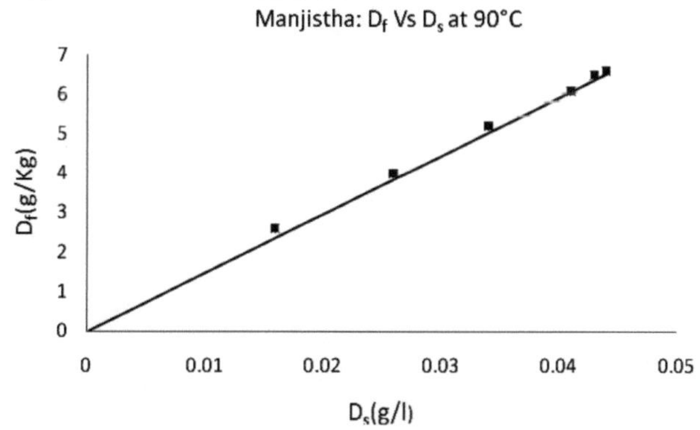

Figure 9.
Plot showing the dyeing isotherm for pre-mordanted jute fabric dyed with Madder/Manjistha at 90°C.

While, **Figure 9** is the Plot between Dye in solution (Ds) Vs Dye in Fibre (Df) at a particular time and temperature (here at 90°C) represent at saturation or equilibrium as corresponding dyeing isotherm.

The chemical affinity ($-\Delta\mu$) for the dye molecule or dyeing affinity for Madder/Rubia/Manjistha towards mordants for pre-mordanted bleached jute fabric when dyed at optimized dyeing conditions for different durations at two different dyeing temperatures (50 and 90°C) is shown in Table 5.2.9. Low but measurable increase in chemical affinity of the said colourant is observed for increase in dyeing temperature from 50–90°C, albeit, higher increase in chemical affinity is expected for increase of dyeing temperature. This moderate value and low increase of chemical affinity for enhancement of dyeing temperature showed that dyeing of bleached and mordanted jute fabric with madder/manjistha do not occur as rapidly as expected and maybe there is low extent of formation of Fibre-Mordant-Dye coordinated complex, while it may be presumed that dyeing occur through weak hydrogen bonding formation in a slower speed. While it is reported in earlier literature [22] that some synergistic effects for application of double pre mordanting with 10% natural potash alum and 10% harda (myrobolan) on cotton before dyeing with madder (Manjistha) due to additional coordinating power of chebulinic acid of harda as a mordanting assistant, facilitates more number of strong and giant bigger complex formation amongst the said fibre (cotton)-mordanting assistants (harda)—metallic mordant (natural alum)—natural dye (madder) to develop higher colour strength and higher Colour fastness to wash as an optimised and better option, which however do not happen in case of dyeing jute fabric with madder/manjistha, after double pre-mordanting with 20% harda (myrobolan) and 20% $Al_2(S0_4)_3$ applied in sequence in this case, may be due to acidity of jute do not allow chebulinic acid of harad (myrobolan) to be attracted/absorbed to jute fibre, as required.

To understand the chemistry of attachment of this particular natural colorant specifically whether the dye molecules from madder or manjistha has been bonded to the fibre-mordant system through pre-dominant H-bonding or through coordinate/chelating complex formation, dyeing isotherm indicate that there is formation of more intermolecular H-bonding between dimeric association of – OH groups of madder component and mordanting assistant like harda (myrobolan) used in double mordant attached through metallic mordant of aluminium sulphate and the jute fibre forming intermolecular H-bonds, and less or no Dye-Mordant Fibre Complex formation occur predominantly as expected. Hence the dyeing isotherm observed is Nernst type (and not Langmuir type) is observed in **Figure 9** like dyeing of non-polar disperse dyes to hydrophobic polyester fibre. However, some metallic chelate formation cannot be excluded fully and need to be explored by FTIR scan etc.

For dyeing of bleached jute after double pre-mordanting with harda (myrobolan) and $Al_2(SO_4)_3$, applied in sequence, heat (enthalpy) of dyeing is found to be positive, showing medium magnitudes of positive values. Thus, this dyeing process may be considered as endothermic and therefore more dye would be adsorbed with increase of dyeing temperature up to equilibrium. In case of double pre-mordanting with harda (myrobolan) and $Al_2(SO_4)_3$ applied in sequence and subsequent dyeing at pH 11.0, K/S value initially increases with increase in dyeing temperature up to 90°C, and above which, the K/S value levelled off. From observed data in **Table 9**, it is indicated that at dyeing temperature between 50–90° C, the ΔH values (required heat of dyeing, as a measure of bond energy/forces of attraction responsible to bind natural dye molecules to the fibre by bridging through the metallic mordant) are always positive in this case but showing lower magnitude of ΔH values within 6.91 to 29.52 kJ/mol. This bond energy values nearly matches with the usual range of bond energy (10–40 kJ/mol) [23] of hydrogen bond formation indicating formation of a weaker dye-fibre bond that has been taken place instead of coordinated co-valent bonds. The +ve sign of ΔH values might have indicated this dyeing process as an endothermic process, which actually occur for hydrogen bond formation between the dye and mordanted fibre. However, metallic mordanting is also essential to increase the attraction of the dye to the fibre in the dye bath during dyeing to increase their chemical affinity and exhaustion of this natural dye towards jute.

$[D]_f$ g/kg		$[D]_s$ g/l		—Δμ kJ/mol		ΔH kJ/mol	ΔS J/mol/°K
at T_1	at T_2	at T_1	at T_2	at T_1	at T_2	for (T_2—T_1)	at T_2
1.5	2.6	0.031	0.016	13.31	18.62	29.52	132.62
2.7	4	0.038	0.026	14.35	18.45	18.83	102.70
3.8	5.2	0.043	0.034	14.93	18.44	13.37	87.61
4.8	6.1	0.047	0.041	15.32	18.35	9.17	75.82
5.5	6.5	0.051	0.043	15.47	18.40	8.23	73.36
5.9	6.6	0.053	0.044	15.55	18.38	7.27	70.65
6.1	6.6	0.054	0.044	15.59	18.38	6.91	69.67

Table 9.
Thermodynamic parameters for dyeing pre-mordanted jute with Madder/Manjistha after double premordanting with harda plus Aluminium sulphate.

Changes in dyeing entropy (ΔS) and dyeing enthalpy (heat of dyeing) are the main indicator of dye absorption and dye fixation force. From observed results in **Table 9**, it is indicated that for different D_f (dye in Fibe) and D_s (dye in solution) values, there is some changes in dyeing entropy at the initial stage of dyeing, with measurable small changes in ΔH values (heat of dyeing), as dyeing time progresses. Df values continues to increase slowly with increase in dyeing time from 30 to 60 min at 90°C in case of said double mordanting system using harda and $Al_2(SO_4)_3$ in pre-mordanting. This slow increase in K/S value, for increase in dyeing time may be due to only physical absorption of dye molecules in fibre by hydrogen bonding with less possibility of Fibre -Mordant-dye co-ordinated complex formation for the dye fixation even on the pre-mordanted fibre, thus without much affecting ΔH and ΔS values.

3.10 Estimation of soil removal efficacy of different detergents used for textiles

For estimation of degree of soiling and soil removal efficiency by standard domestic laundering by selective detergent [24], first the clean white or light coloured fabrics are to be artificially soiled under standard conditions by dipping and running the clean fabric under an oil in water emulsion with water+ coconut oil/carbon tetrachloride with addition of recommended dosages of graphite powder or carbon black powder and the changes in reflectance value after artificial soiling gives degree of soiling as depicted in the following Eq. 25;

$$\text{Degree of Soiling (\%Soiling)} = \frac{R_0 - Rs}{Rs} \times 100 \tag{25}$$

Where $\mathbf{R_0}$ is the Initial Reflectance value of clean (unsoiled) white/light coloured fabric and **Rs** is the Reflectance value of artificially soiled white or light coloured fabric.

Further, estimation of soil removal efficacy % of any detergent, can be similarly calculated by change of Reflectance of corresponding soiled fabric sample before and after washing at specified standard conditions in launder-o-meter, represented by following Eq. 26:

$$\begin{array}{c}\text{Degree of Soil removal efficacy (\%)} \\ \text{Or Percent soil removal Efficiency}\end{array} = \frac{R_L - Rs}{R_0 - Rs} \times 100 \tag{26}$$

where **Rs** is the Reflectance value of artificially soiled white or light coloured fabric before laundering and $\mathbf{R_L}$ is Reflectance values of the standard soiled fabric after Laundering for given numbers of cycles of wash under specified washing conditions of domestic wash under lauder-o-meter. Also, to determine degree of soil redeposition %, AATCC Test Method 151 can be used to estimate the degree of soil redeposition likely to occur during laundering as soil removal efficiency is never 100% and gradual redeposition of soil on fabrics under wash always occurs. The fabrics to be tested are exposed to initially to a standard soiling method (preferably taking fabric swatches with both dry soiling followed by fabric pretreated with a standard oily soil) and then subjected to laundering to determine both soil removal efficacy and soil redeposition during a laundering simulated with a standard domestic wash with selective detergent. The change in reflectance of the fabric before and after laundering for the soiled fabric under testing is an indication of the % soil redeposition potential of the fabric as well as soil removal efficacy percent of the corresponding detergent.

4. Concluding remarks

The application of above said colorimetric analysis with few case studies for textile industry are a small glimpse only considering this vast subject of colorimetry and hence, this can be applied in makeshift way to other different industry as well. In the colorimetric analysis, besides conventional old model of colorimeter (which is almost abandoned) UV VIS absorbance spectrophotometer and UV VIS Reflectance spectrophotometer, both took major role for colorimetric analyses of all types of Liquid and solid coloured samples used in textile industry, paint industry, food industry, chemical industry, cosmetic industry, pharmaceutical industry etc., where colour information could be obtained with different type of sensor/detector to quantify the colour variation in different colour spaces such as CIE L*a*b* colour space and other recent few more colour space used such as CIE-LUV, RGB, CMC etc., Besides the conventional approaches of colorimetric analysis, non-conventional approaches are now being applied on liquid samples for detection of chlorine in water, to check ripeness estimation of different fruits, to check colour differences in blood to determine blood shading date (or age) for forensic purpose, to determine efficacy of UV active agents like Bluing agents or optical brighteners/ UV absorbers used in textile industry etc., where quantification of required colour parameters are calculated using analytical formulas extracted from different colour space concepts defined and measured using UV VIS absorbance spectrophotometer and UV VIS Reflectance spectrophotometer. Presently Portable Reflectance spectrophotometer are the industry's major choice due to its handy use and carrying capability from one place to other.

As an alternative to UV-VIS spectrophotometric analysis, colorimetry is also widely used in many applications including food allergen testing, albumin testing in urine analysis, blood analysis, pH quantification and water monitoring in different industry.

Over the last decade, scientist has made possible that smartphones may also be used in a variety of scientific fields as spectrometers or as colorimeters, if provided with optical sensor. Smartphone optical spectrometers uses the wavelength scan components, which give spectral information at 400 to 700 nm for the collimated light from the optical source which is dispersed after interaction with samples and corresponding results are recorded. The colour spectrum image of the sample taken in a smart phone is transformed into various colour spaces for the extraction of quantitative colour data. The wavelength of the spectrum generally changes between 400 and 700 nm because of the optical filters set in front of the camera in the manufacturing process which serves the purpose of using this Spectral information in many applications from smart phone.

Smartphone-based spectrometer and colorimetry have been gaining popularity and current relevance due to the widespread advances of these type of small sized and multipurpose smart devices with increasing computational and spectral recording power having relatively low cost and portable designs with very much user-friendly interfaces, and compatibility with data acquisition and processing facility. They find applications in interdisciplinary fields, including but not limited to textiles or paints or pharmaceutical industry, agriculture industry chemical industry and biological and medical purposes too.

Author details

Ashis Kumar Samanta
Department of Jute and Fibre Technology, University of Calcutta, Kolkata, West Bengal, India

*Address all correspondence to: ijtaksamanta@hotmail.com

References

[1] CIE Publ. International Lighting Vocabulary. Vol. No. 17.4. 1976 & 1987. pp. 1-5

[2] Shah HS, Gandhi RS. Instrumental Colour Measurements & Computer Aided Colour Matching for Textiles. Amhedabad, India: Mahajan Book Distributers; 1990. pp. 76-116

[3] Berns RS. Billmeyer and Saltzman's, Principles of Colour Technology. 3rd ed. New York, USA: John Wiley & Sons; 2000

[4] Ohta N, Robertson AR. Colorimetry-Fundamentals and Applications. New York & England: John Wiley & Sons; 2005

[5] McLaren K. Colour Science of Dyes and Pigments. 2nd ed. Bristol: Adam Hilger; 1986. pp. 50-87

[6] Hunt RWG. Measuring Colour. 3rd ed. Kingston-upon Thames: Fountain Press; 1998

[7] Pubalina S. Fundementals and applications of of Computer aided Colour Match Prediction (CCMP) System. Latest Trends in Textile and Fashion Design Journal. 2018; **2**(5):1-10

[8] Ingamells W. Colour for Textiles-A user's handbook. Bradford, UK: Society of Dyers and Colourists; 1993. pp. 138-157

[9] IS standards-17085: 2019: Textile Dyestuffs-Rubia (Maddar)-Identification, 2019.

[10] ISO/Standards-22195-1-2019 (E). Textiles-Determination of index ingredient from coloured textiles — Part 1. Madder; 2019

[11] ISO-2470-1977 (E) and ISO-2469, 1977. Determination of Brightness. Switzerland: International Organization for Standardization (ISO); 1977. pp. 1-3

[12] Samanta AK, Agarwal P, Singhee D, Dutta S. Application of single and mixtures of red sandalwood and other natural dyes for dyeing of jute fabric: Studies on colour parameters/ colour fastness and compatibility. Journal of Textile Institute. 2009; **100**(7):565-587

[13] Oulton DP, Porat I. Control of Colour by Using Measurement and Feedback. Journal of the Textile Institute. 1992;**83**(3):454-461

[14] Randall D, Stutts T. Optimizing calibration dyeings for computer color matching. Amer. Dyestuff. Rep. 1988; **28**:44-46

[15] Bruce MA, Benson L, Oulton DP, Hogg M, Wilson J. Optimising product colour. In: Review of Progress in Coloration, Society of Dyers and Colourists. Vol. 31. Bradford; 2001. pp. 29-35

[16] Shukla SR, Dhuri SS. Assessing the compatibility of disperse dye mixtures by the use of colour coordinates. Journal of the Society of Dyers and Colourists. 1992;**108**:139-144

[17] Haleh Khalili and Seyed Hossein Amirshahi, A Novel Method for Determination of Compatibility of Dyes by Means of Principal Component Analysis, Color Research & Application. August 2010;**35**(4):313-318

[18] Kumar SA, Agarwal P, Datta S. Studies on Color-Strength-Related Parameters and Compatibility for Dyeing of Cotton Fabric with Binary Mixtures of Jackfruit Wood and Other Natural Dyes. Journal of Natural Fibers. 2009;**6**(1):27-45

[19] Datye Keshav V, Mishra S. Compatibility of Dye Mixtures. Journal of the Society of Dyers and Colourists. 1984;**100**:334-339

[20] Samanta AK. A Konar, Nilendu Sekhar Bhoumick and A N Ray, Studies on Compatibility of Selective Direct Dyes for Dyeing of Jute Fabric. Ind. J of Fibre and Text Res. 2019;**44**:98-106

[21] Samanta AK. On-line colour control (for textile dyeing). Ind. Text. Journal. 1992;**102**:70-76

[22] Samanta A K, Bio-Dyes, Bio Mordants and Bio-Finishes: Scientific Analysis for Its Application on Textiles – a book chapter In: "Chemistry and Technology of Natural and Synthetic Dyes and Pigments", Ed. by A. K. Samanta (Chief Editor) and Naseer Awwad and Hamed Majdooa Algarni (Jt Editors), on line and Print version. London UK: Published by In Tech Open; 2020. pp. 3-37

[23] Samanta AK, Konar A, Sarmishtha Chakrabarti (Mukherjee), Siddhatha Datta. Effect of different Mordants, Extraction Conditions and Dyeing Process Variables on colour interaction Parameters and Colour Fastness properties in dyeing of Jute Fabric with Manjistha, a Natural Dye, Journal of Institution of Engineers (India)-Textile Engg. 2010;**91**:7-15

[24] Samanta AK, Mitra S, Singhee D, Parikh S. Efficacy of selective surfactant/detergent as washing agents on soiled white/dyed cotton fabrics. Ind. J Fibre Text Res. 2004;**29**:223-232

Chapter 2

UV-Visible Spectroscopy for Colorimetric Applications

Sonia Karuppaiah and Manikandan Krishnan

Abstract

UV-visible spectroscopy is an interpretive skill that amplitude the variety of different wavelengths of UV or visible light, which are captivated by or transferred via a pattern new assessment to an implication or blank constituent. This asset is encouraged by way of the pattern combination, doubtlessly subject to network on what is within the representative and at what attention. Because this spectroscopy execution confides on the control of mild. Therefore, illuminate can be described by its wavelength, which can be useful in UV-visible spectroscopy to analyse or identify different substances.

Keywords: detectors, filters, monochromators, sources, UV-visible spectroscopy

1. Introduction

The analytical chemistry is based on the quality of colour in coloured solution, we observe the colour, the colour's depth, or intensity. These observations led to the technique called colorimetry, the colour of a solution identify species while the intensity of the colour depends on identifying the concentration of the species present. The important and sensitive colour tests have been developed for the detection and determination of a wide range of chemical species, both inorganic and organic in nature, this used the development of visible and ultraviolet spectrometer [1].

The wavelength range of UV radiation starts at 400 nm, the blue end of visible light, and ends at 200 nm. The radiation has sufficient energy to excite electrons. When light passes through the solution and emerges as red light, then the solution is red. Because the solution has allowed the red component of white light to pass through, whereas if the solution has led the red component of white light to pass through because it has absorbed the complementary colours, yellow and blue [2].

If the solution has more concentration, more yellow and blue light is absorbed, and more intensely red solution appears to the eye. There is a difficulty in comparing the intensity of the two colours. The wavelength range of UV radiation starts at the end of visible light of 400 nm and ends at 800 nm [3]. The atoms or molecules have sufficient energy to excite valence electrons. Visible light starts the wavelength from 800 to 400 nm.

2. Theory

2.1 Electronic excitation in molecules

The atoms are held strongly by sharing electrons in a molecule. The electron in a molecule moves in molecular orbitals at discrete energy levels. When the energy of the electrons is at a minimum, the molecules are in the lowest energy state or ground state. The molecules can absorb radiation and move to a higher energy state or excited state. The movement of electrons from a higher energy state is called electronic excitation [4]. The frequency captivates or effuse by a molecule and the power is related by, $\Delta E = h\gamma$. The amount of energy required is based upon the variation in energy linking the ground state E_0 and the excited state E_1 of the electron. It is stated as $\Delta E = E_1 - E_0 = h\gamma$

where, E_1 is the energy of the excited state.

E_0 is the energy of the ground state.

The full strength of a molecule is the same as the sum of electronic, vibrational, and rotational electricity. The importance of the energies decreases inside the following order: E_{elec}, E_{vib}, and E_{rot}. Ultraviolet energy is computed, the assimilation spectrum arising from a single electronic transition must contain a single discrete line. However, an awesome line is not obtained because digital absorption is superimposed upon rotational and vibrational sublevels. Suppose of complex molecules in conjugation with an excess of two atoms, discrete bands merge to bring about broad absorption bands or "band envelops" [5]. Three distinct types of electrons are involved in organic molecules. They are as follows:

i. σ electrons: Electrons associated with the single bonds are known as σ electrons. Electrons are involved in saturated bonds, such as those between carbon and hydrogen-like C-C, C-H, O-H. As the amount of energy required to excite electrons in σ bonds is much more than that produced by UV light. Example: Hexane C_6H_{14}.

ii. π electrons: Electrons are involved in a double and triple bond that is involved in unsaturated hydrocarbon-like alkenes, alkynes, conjugated olefins, and aromatic compounds.

iii. n electrons: Electrons that are not involved in bonding between atoms or molecules. Organic compounds containing nitrogen, oxygen, sulphur or, halogens.

2.2 Electronic transition in organic molecule

A rule to predict how molecules undergo a transition is given by Quantum mechanics. Some transitions are "allowed" while others are "Forbidden."

i. **$\sigma \rightarrow \sigma^*$:** Orbitals are conserved; therefore, two molecular orbitals are formed, a sigma bonding orbital and a higher energy sigma antibonding orbital. The antibonding orbital is denoted by σ^*. The energy difference between σ and σ^* is equal, denoted by ΔE. The compounds in which all valence electrons are involved in the single bond formation such as saturated hydrocarbon show absorption in far UV radiation below 190 nm.

ii. n $\rightarrow$ **σ^*:** the transition takes place in saturated compounds containing one hetero atom with unshared pair of electrons (n electron). The compounds which undergo these transitions are saturated halides, alcohols, ethers, aldehydes, ketones, amines, etc. This transition requires less energy. In saturated alkyl halides, the energy required for such a transition decreases with the increase in the size of the halogen atom.

In methyl chloride and methyl iodide due to the electronegativity of chlorine atom, the n electrons on chlorine atom are comparatively difficult to excite, whereas the methyl iodide is 258 nm as n electrons on iodide atom are loosely bound.

iii. $\pi \rightarrow \pi^*$: The transitions occur in unsaturated compounds that contain double and triple bonds and in aromatics. The excitations of π electron require smaller energy and hence transitions of this type occur at a longer wavelength. π electron of a double bond is excited to π^* orbital. The compounds that undergo are alkenes, alkynes, carbonyl compounds, cyanides, azo compounds, etc.

iv. $n \rightarrow \pi^*$: the compounds with a functional group such as C=O, C=S, C=N undergo $n \rightarrow \pi^*$. This type of transition requires the least amount of energy. The compounds like nitrogen, oxygen, Sulphur, halogen atom especially Br and I in UV/visible region undergo transition with nonbonded electrons [6]. The electronic transitions are shown in **Figure 1**.

2.2.1 Beer's and Lambert's law

There are two laws related to the absorption of radiation [7].

$$I = Ia + It$$

I = Intensity of incident light.
Ia = Intensity of absorbed light.
It = Intensity of transmitted light.

2.2.2 Beer's law

The intensity of a beam of monochromatic light drops exponentially with expanding in the concentration of absorbing species arithmetically.

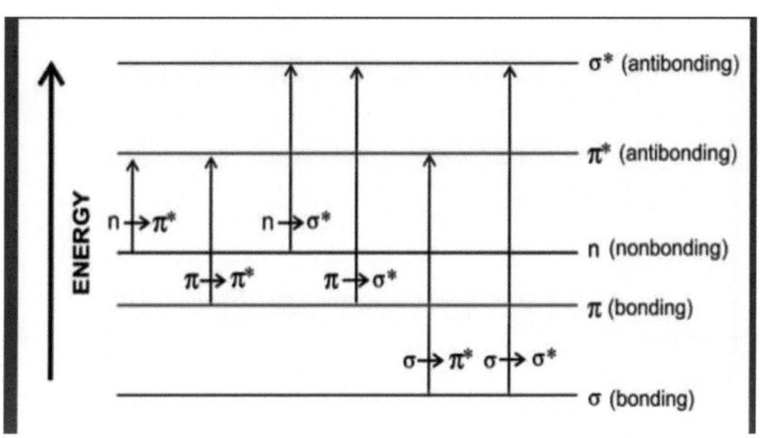

Figure 1.
The electronic transition.

$$-\frac{dI}{dc} \propto I \text{ (the decline in the intensity of incident light, with concentration, C is proportional to the strength of incident light, I)}$$

$$-\frac{dI}{dc} = K\,I \text{ (eliminate and introducing constant proportionality K)}$$

$$-\frac{dI}{dc} = K\,dc \text{ (rearranging terms)}$$

$$-\ln I = Kc + b \text{ (on integration)} \quad (1)$$

When concentration, C = 0, there is no absorbance $I = I_0$.
Exchange in Eq. (1).

$$-\ln I_0 = K \times 0 + b$$

$$-\ln I_0 = b$$

Substitute the value of $-\ln I_0 = b$ in Eq. (1).

$$-\ln I = Kc - \ln I_0$$

$$-\ln I_0 - \ln I = Kc$$

$$\ln I_0/I = Kc \text{ (Since } \log A - \log B = \log A/B) \quad (2)$$

$$I_0/I = e^{Kc} \text{ (separate natural logarithm)}$$

$$I/I_0 = e^{-Kc} \text{ (reversed on both sides)}$$

2.2.3 Lambert's law

The rate of decrease of intensity (monochromatic light) with the thickness of the medium is directly proportional to the intensity of incident light.

$$-\frac{dI}{dt} \propto I \text{ (the decline in the intensity of incident light, with concentration, C is proportional to the intensity of incident light, I)}$$

$$-\frac{dI}{dt} = K\,I \text{ (separate and introducing constant proportionality K)}$$

$$-\frac{dI}{I} = K\,dt \text{ (reposition terms)}$$

$$-\ln I = Kt + b \text{ (on integration)} \quad (3)$$

When concentration, t = 0, existent is never absorbance $I = I_0$.
Substituting in Eq. (3).

$$-\ln I_0 = K \times 0 + b$$

$$-\ln I_0 = b$$

Substitute the rate of $-\ln I_0 = b$ in Eq. (1).

$$\begin{aligned} &-\ln \mathrm{I} = \mathrm{Kt} - \ln \mathrm{I}_0 \\ &-\ln \mathrm{I}_0 - \ln \mathrm{I} = \mathrm{Kt} \\ &\ln \mathrm{I}_0/\mathrm{I} = \mathrm{Kt}\ (\text{Since logA} - \log \mathrm{B} = \log \mathrm{A/B}) \\ &\mathrm{I}_0/\mathrm{I} = e^{Kt}\ (\text{removing natural logarithm}) \\ &\mathrm{I}/\mathrm{I}_0 = e^{-Kt}\ (\text{Inverse on bilateral}) \end{aligned} \tag{4}$$

Combine and equate Eqs. (3) and (4)

$$\begin{gathered} \mathrm{I}/\mathrm{I}_0 = e^{-Kct} \\ \mathrm{I} = Ioe^{-Kct} \\ \mathrm{I} = Io10^{-Kct}\ (\text{Converting natural logarithm to base 10\&K} = \mathrm{K} \times (0.4343) \\ \mathrm{I}/Io = 10^{-Kct}\ (\text{reposition terms}) \\ \mathrm{Io/I} = 10^{Kct}\ (\text{reverse on both sides}) \\ \text{Log Io/I} = \mathrm{Kct}\ (\text{Taking log on both sides}) \end{gathered} \tag{5}$$

Transmittance (T) = I/Io and Absorbance (A) = log 1/T.
Hence $\mathrm{A} = \log \frac{1}{I/Io.}$

$$\mathrm{A} = \log \mathrm{Io/I} \tag{6}$$

Substitutes Eq. (6) in Eq. (5)

$$\begin{aligned} &\mathrm{A} = \mathrm{KCt}\ (\text{Instead of K, we write } \varepsilon) \\ &\mathrm{A} = \varepsilon \mathrm{ct} \end{aligned}$$

Where, A = Absorbance or optical density or extinction coefficient
ε = molecular extinction coefficient
C = Concentration of drug (mmol/lit)
T = pathlength (1 cm)
ε can also expressed as

$$\varepsilon = E\sum_{1cm}^{1\%} xMolecularweight/10$$

Where $E\sum_{1cm}^{1\%}$ means the absorbance of 1% W/V solution using a path length of 1 cm.

2.3 Deviation from Beer's and Lambert's law

Beer and Lambert's law is found to be obeyed by the system if a straight line passes through the origin and a graph is plotted between absorbance and concentration.

But there is always a deviation from the linear relationship between the absorbance and concentration particularly at higher concentration, and hence the absorption curve changes with the change in concentration of the solution. The deviation may be positive or negative, if the resulting curve is concave upwards it is called positive deviation. If the resulting curve is concave downwards it is called negative deviation, which is depicted in **Figure 2** [8].

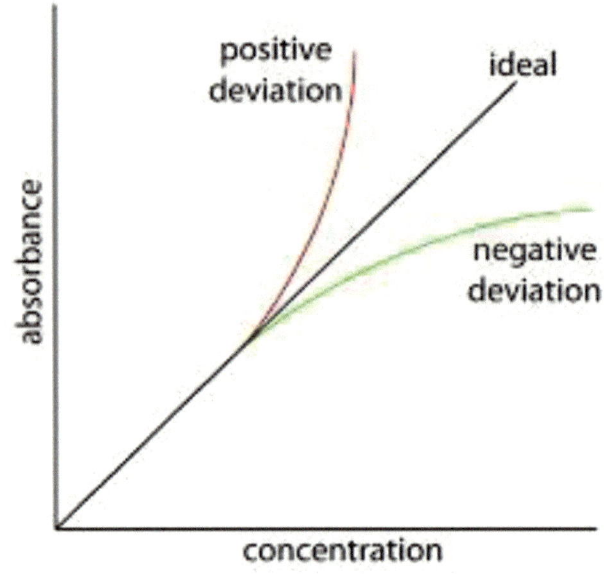

Figure 2.
Deviation from Beers & Lamberts law.

2.4 The reason for changing deviation from Beer's law

2.4.1 Instrumental deviation

Factors like stray radiation improper slit width, fluctuations in single and when monochromatic light is not used.

2.4.2 Physiochemical changes in solution

i. The law does not hold if the substance ionises, dissociates, or associates in solution. Since the nature of the ionised species in solution varies with the concentration.

Example: Benzoic acid in benzene is associated to form dimer and hence deviation occurs.

i. Potassium dichromate in high concentration exists as orange solution (λmax −450 nm). But on dilution, dichromate ions are dissociated into chromate ions, which are yellow-coloured (λmax −410 nm).

ii. When sufficient time is not allowed for making absorbance measurement or when the reading is made when the colour has faded away due to instability of colour, deviation can occur due to incomplete reaction.

iii. If a solute forms complexes, the composition and extent of complexation depend upon the concentration.

iv. A large number of electrolytes may shift the λmax and change the extinction coefficient.

v. If the change in concentration causes significant alterations in the refractive index, then deviations from the law.

vi. Changes in pH with a change in concentration of solute may cause deviation.

vii. The presence of impurities that fluoresce or absorb at the required absorption, the wavelength may cause deviation.

3. Effects of solvents UV spectra

Chromophore: the term chromophore is used to denote a functional group or presence of some structural feature that gives colour to a compound [9].

Example: Nitro group is a chromophore because its presence in a compound gives the yellow colour to the compound. It can be defined as any group which exhibits absorption of electromagnetic radiation in the visible or ultraviolet region. It may or may not impart any colour to the compound. Some of the important chromophores are ethylenic, acetylenic, carbonyls, acids, esters, nitrile group, etc.

There are two types of chromophores. The chromophore in which the group contains π electrons and they undergo $n \rightarrow \pi^*$ transitions, the compounds like ethylene, acetylene, etc.

The other type of chromophore contains both π electrons and n (non-bonding) electrons. This type of chromophore undergoes two types of transitions, $\pi \rightarrow \pi^*$ and $n \rightarrow \pi^*$ and examples include carbonyls, nitriles, azo compounds, and nitro compounds.

3.1 Changes in position and intensity of absorption

For isolated chromophore groups such as >C=C < and -C ≡ C-, absorption takes place in the far ultraviolet region which cannot be easily studied.

But the role of absorption is maximum and the intensity of absorption can be edited in exceptional approaches by some structural adjustments or change of solvent.

3.1.1 Bathochromic shift or redshift

It involves the shift of absorption most in the direction of longer wavelength because of the presence of certain groups such as OH and NH_2 called auxochromes or by change of solvent. A Bathochromic shift is also produced when two or more chromophores are present in conjugation in the molecule.

Example: Ethylene shows $\pi \rightarrow \pi^*$ transition at 170 nm, whereas 1,3 -butadiene (where two double bonds are in conjugation) shows λmax at 217 nm.

3.1.2 Hypsochromic or blue shift

The shift of absorption maximum towards shorter wavelength and may be by the removal of conjugation or by change of solvent. The absorption shift towards a shorter wavelength is also called the blue shift.

Example: Aniline shows maximum absorption at 280 nm, because the pair of electrons on the nitrogen atom is in conjugation with the π bond system of the benzene ring. In acidic solution, a blue shift is caused and absorption takes place at a shorter wavelength 200 nm. The electron pair is no longer present and hence conjugation is removed.

Hyperchromic effect: The effect is due to an increase in the intensity of absorption and it is brought about by the introduction of an auxochrome.

Hypochromic effect: It involves a decrease in the intensity of absorption and is brought by groups that are able to distort the geometry of the molecule.

Auxochrome: It is a group that itself does not act as a chromophore but when attached to a chromophore it shifts the adsorption maximum towards a longer wavelength along with an increase in the intensity of absorption.

3.2 Instrumentation

The various components of a UV-VIS spectrophotometer are as follows [3]:

1. Radiation source
2. Monochromators
3. Detector
4. Recording system
5. Sample cells
6. Matched cells
7. Power supply

4. Radiation source

In UV-VIS spectrophotometer, the normally pre-owned radiation is preferred to assets the hydrogen or deuterium lamps, the xenon discharge lamps, and mercury arcs. In all the assets, agitation is carried out by means of transient electrons through gasoline and those impacts in the midst of electron and gas molecules may bring about digital, vibrational, and rotational elation in the fume's particle [10].

The following are requirements of a radiation source:

1. It must be stable.
2. It must be sufficient intensity for the transmitted energy to be detected at the end of the optical path.
3. It must supply continuous radiation over the entire wavelength.

4.1 Tungsten lamp

The function is similar to an electric light bulb. It is a tungsten filament heated electrically to white heat. The structure is depicted in **Figure 3**.

4.1.1 Disadvantage

1. The intensity of radiation at a short wavelength < 350 nm is small.
2. To maintain a constant intensity, the electrical current to the lamp must be controlled.

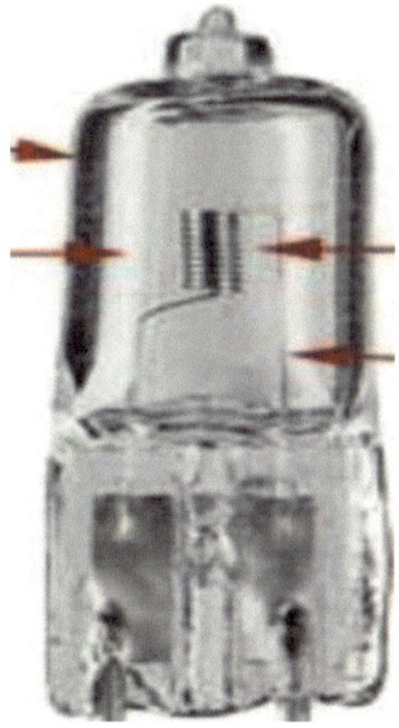

Figure 3.
Tungsten lamp.

4.1.2 Advantage

The lamps are generally stable, robust, and easy to use.

5. Hydrogen discharge lamps

Hydrogen gas is stored under relatively high pressure. When an electric discharge is passed through the lamp, excited hydrogen molecules will be produced which emit UV radiations. Hydrogen lamps cover the range of 3500–1200A°. These lamps are stable, robust, and widely used.

Hydrogen discharge lamp consists of hydrogen gas under relatively high pressure through which there is an electrical discharge. The hydrogen molecules are excited electrically and emit UV radiation. The high pressure brings many collisions between the hydrogen molecules, resulting in pressure broadening. This causes the hydrogen to emit a continuous broadband rather than a simple hydrogen line spectrum. It is stable, robust, and widely used. It is more expensive is the disadvantage.it is depicted in **Figure 4**.

6. Deuterium lamp

It is used in place of hydrogen, the intensity of radiation emitted is 3–5 times the intensity of a hydrogen lamp of comparable design. It is more expensive than a hydrogen lamp. But it is used when high intensity is required. It is represented in **Figure 5** [11].

7. Xenon discharge lamp

Xenon gas is stored under pressure in the range of 10–30 atmospheres. The xenon lamp possesses two tungsten electrodes separated by about 8 mm. When an

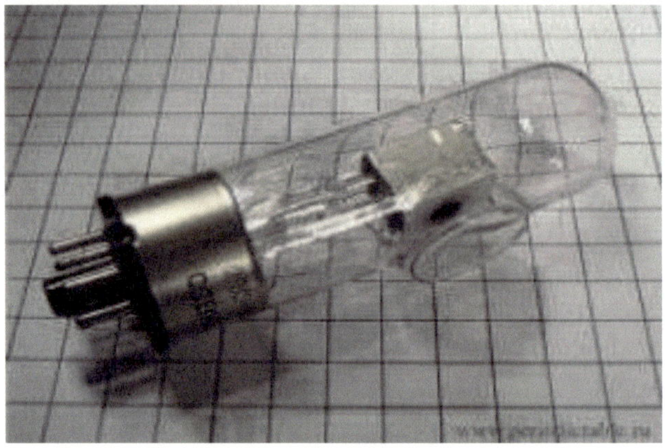

Figure 4.
Hydrogen discharge lamp.

Figure 5.
Deuterium lamp.

Figure 6.
Xenon discharge lamp.

intense arc is formed between two tungsten ultraviolet light is produced. The structure is depicted in **Figure 6**.

8. Mercury arc

Mercury vapour under high pressure, the excitation of mercury atoms is done by electric discharge. It is not suitable for continuous spectral studies because of the presence of sharp lines or bands. It is depicted in **Figure** 7.

9. Monochromators

The monochromator is used to disperse the radiation. The essential elements of a monochromator are:

- Entrance Slit (to get narrow source)

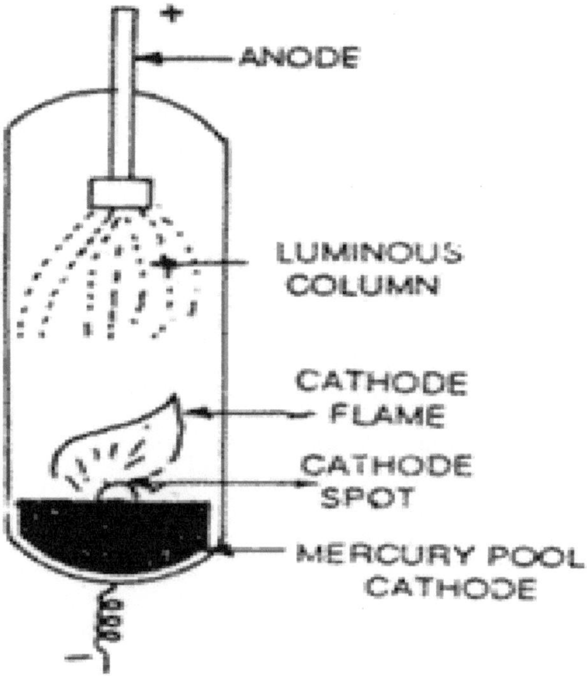

Figure 7.
Mercury arc.

- Collimator (to render light parallel)
- Grating or Prism (to disperse radiation)
- Collimator (to reform the images of entrance slit)
- Exit slit (to fall on sample cell)

Monochromators are better and more efficient than filters in converting polychromatic light or heterochromatic light into monochromatic light. The structure is depicted in **Figure 8**.

9.1 Prism

The prism disperses the light radiation into individual colours or wavelengths. These are found in expensive instruments. The bandpass is lower than that of filters and hence it has better resolution and is depicted in **Figure 9**.

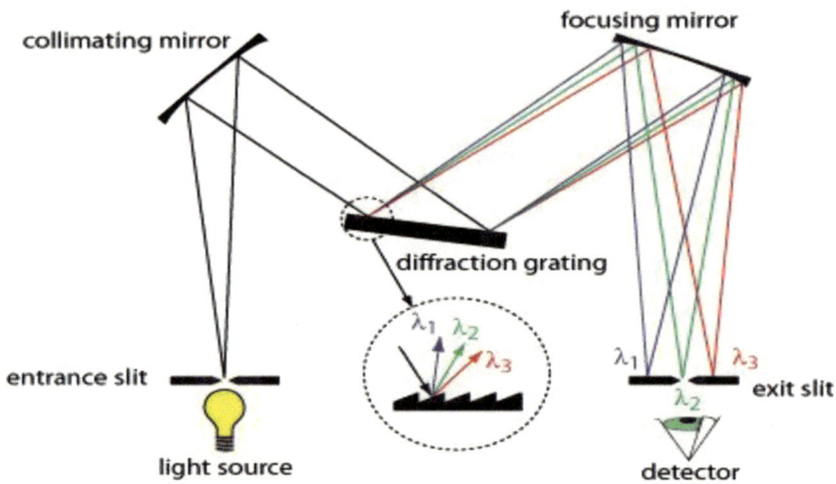

Figure 8.
Monochromators.

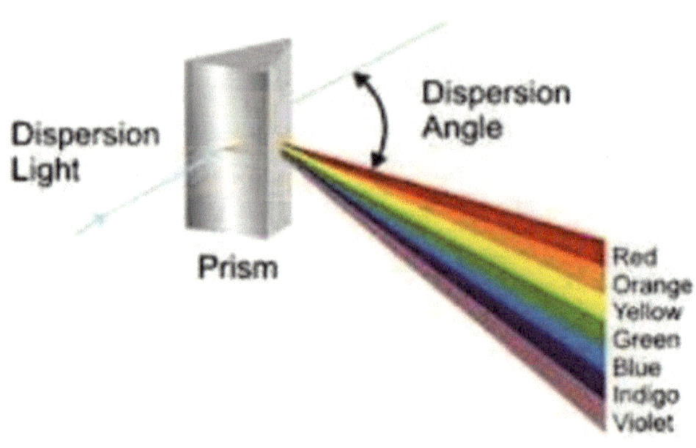

Figure 9.
Prism.

The two types of the prism are:

1. Refractive

2. Reflective

They undergo dispersion giving wavelengths that do not overlap and the disadvantage is they give non-linear dispersion.

9.1.1 Refractive type

The sources of light, through the entrance slit falls on a collimator. The parallel radiations from the collimator are dispersed into distinctive colorations or wavelength, and through the use of any other collimator, the pix of the front slit is reformed. The reformed ones will be both violet, indigo, blue, green, yellow, orange, or pink. The desired radiation on go-out slit may be decided on with the aid of rotating the prism or by way of preserving the prism stationary and transferring the exit slit which is depicted in **Figure 10**.

9.1.2 Reflective type

The dispersed radiation gets reflected and can be collected on the same side as the source of light.

9.1.2.1 Grating

Grating are the most efficient ones in converting a polychromatic to monochromatic light. Two types of the grating are diffraction and transmission.

9.1.2.2 Diffraction grating

A grating consists of a large number of parallel lines (grooves) ruled on a highly polished surface such as alumina, generally, 15,000–30,000 lines per square inch are drawn. When light rays have impinged on the grating, its grooves act as

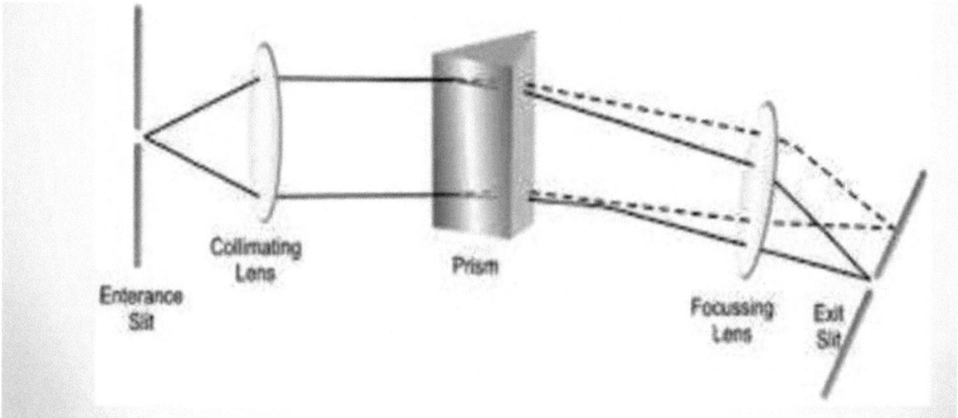

Figure 10.
Refractive prism.

scattering centres for light rays. The light is diffracted or reinforcement takes place. Grating are difficult to be prepared. The replica grating is prepared from an original grating. This is done by coating the original grating with a film of an epoxy resin, which after setting is removal to yield a replica (**Figure 11**).

$$m\lambda = b\,(\text{Sin i} \pm \text{Sin r})$$

λ = wavelength of light produced
b = grating spacing
i = angle of incidence
r = angle of reflection
m = order

9.1.2.3 Transmission grating

Refraction takes place instead of reflection. The wavelength of radiation produced by transmission grating can be expressed by the following equation, the structure is depicted in **Figure 12**.

$$\lambda = \left(\frac{d\sin\Theta}{m}\right)$$

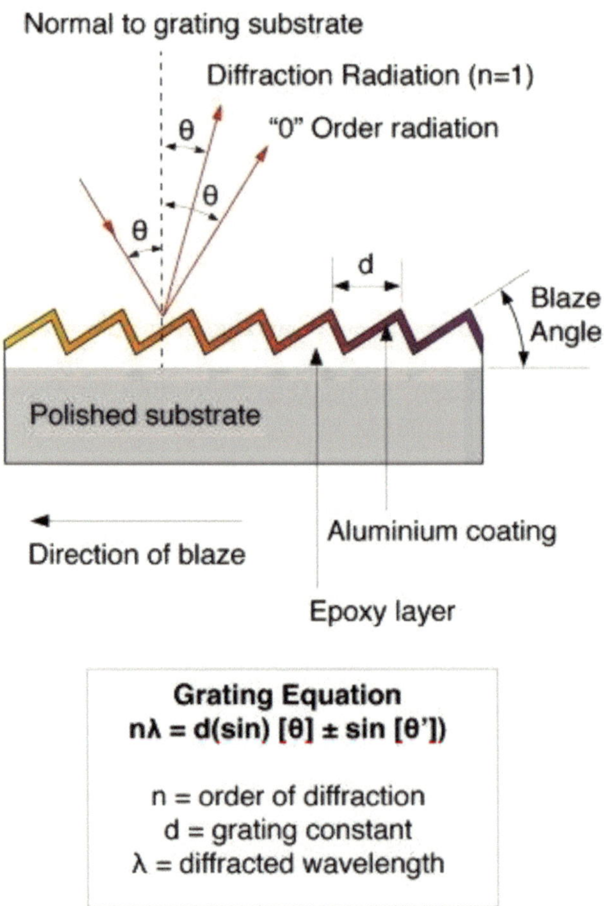

Figure 11.
Diffraction grating.

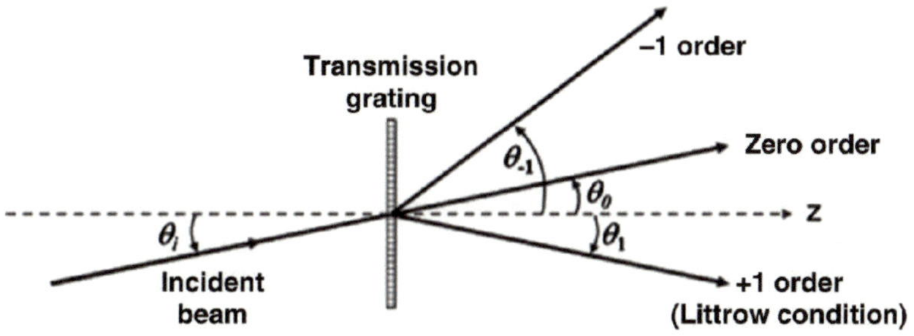

Figure 12.
Transmission grating.

λ = wavelength of radiation
d = 1/lines per cm
m = order no. (0, 1, 2, 3, ... etc.)
ϴ = angle of deflection
d = 1/2000 = 0.0005 = 5×10^{-4}
ϴ = 6.89°

10. Detector

Whilst a radiation is passed via a pattern cellular, part of its miles being absorbed by means of the pattern solution and rest is being transmitted. The transmitted radiation falls on the detector and the intensity of absorbed radiation can be decided.

10.1 Barrier layer or photovoltaic cell

The barrier mobile includes a semiconductor, consisting of Selenium that is deposited on a sturdy steel base, inclusive of iron. A completely skinny sheet of silvery or aurelia is stammer ended the surface of the semiconductor to behave as collector electrode. The emission falling at the floor yield electron at the selenium silver interfaces. A barrier exists between the selenium and iron, which rule out the electrons against streaming into iron. The electrons are collected on the silver surfaces. The buildup of electrons on the silver surfaces produces an electric voltage distinction between the silver surfaces and the base of the mobile.

If the peripheral circuit secures a low resistance, a photocurrent will glide, that is precisely equivalent to the intensity of the incident radiation beam. It is holed directly to micrometre or galvanometer to read its output (**Figure 13**) [12].

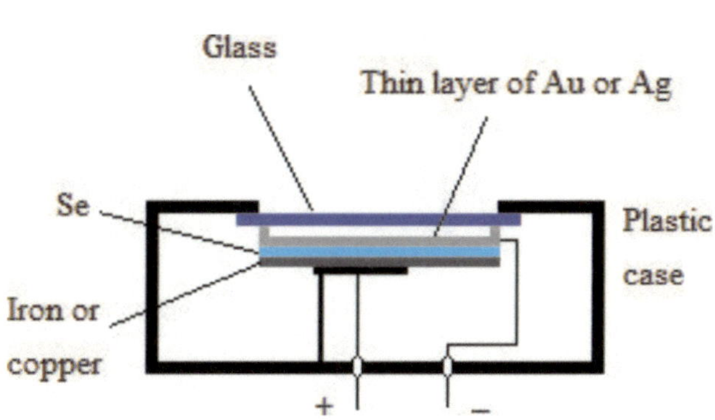

Figure 13.
Barrier layer or photovoltaic cell.

10.2 Phototubes or photoemissive cell

It consists of a high-sensitive cathode in the form of a half-cylinder of metal is contained in an evacuated tube. The inside surface of the photocell is coated with a light touchy layer. While the mild is incident upon a photocell, the floor coating emits electron. Those are attracted and amassed by an anode. The modern-day, that is created among the cathode and anode, is seemed as a measure of radiation falling at the detector. A phototubes is greater touchy than photovoltaic cellular due to the fact excessive diploma of amplification can be used (**Figure 14**) [13].

10.3 Photomultiplier tubes

A photomultiplier tube is a combination of a photodiode and an electron-multiplying amplifier. A photomultiplier tube consists of an evacuated tube that contains one photo-cathode and 9–16 electrodes referred to as dynodes. The surface of each dynode is Be-Cu, Cs-Sb.

While radiation falls on a metallic floor of a photocathode, it emits electrons. The electrons are attracted towards the primary dynode that is kept of a fine voltage. While the electron strikes the primary dynode which is saved at a wonderful voltage. Whilst the electron strikes the first dynode, extra electrons are emitted with the aid of the floor of the dynode; these emitted electrons are then attracted via a second dynode, wherein comparable sort of electron emission take area. The technique is repeated over all of the dynodes gift in the photomultiplier tube until a bath of electrons reaches the collector. The range of electrons accomplishing the collector is the degree of the depth of light falling at the detector (**Figure 15**).

11. Recording system

The signal from the detector is finally by the recording system. The recording is done by recorder pen.

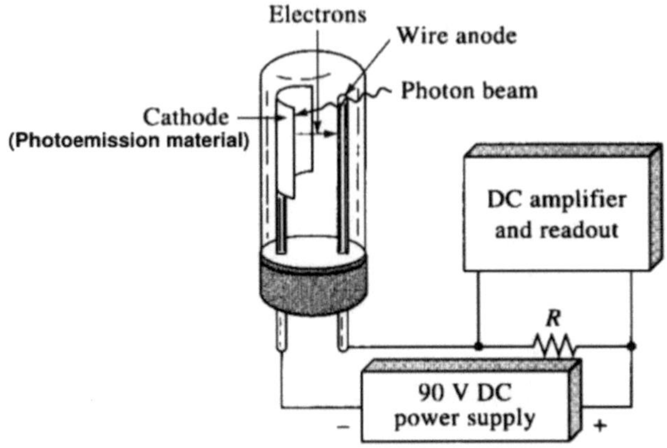

Figure 14.
Phototubes or photoemissive cell.

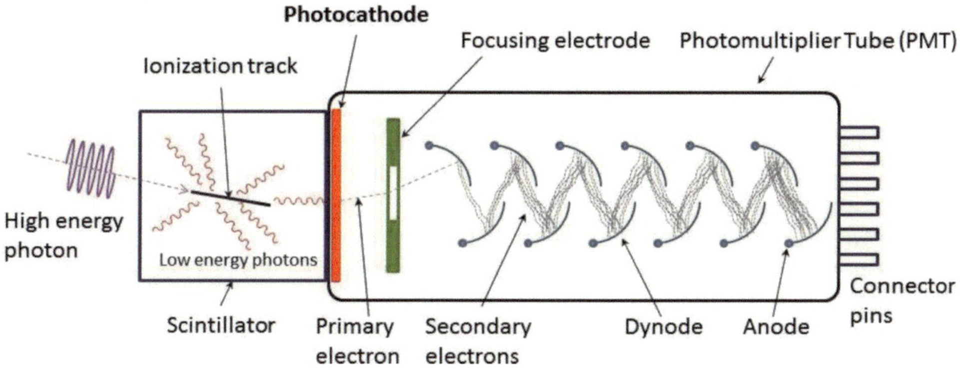

Figure 15.
Photomultiplier tubes.

12. Sample cells

The cells must contain:

i. Uniform in construction

ii. Material used for construction should be inert to solvents.

iii. They must transmit light of the wavelength.

The commonly used cells are made of quartz or fused silica (**Figure 16**) [13].

13. Matched cells

When double beam is used, two cells are needed, one for the reference and one for the samples. It is normal for the absorption by the cells to differ slightly. This causes a small error and can lead to analytical error, so matched cells are used. When the same matched cells are used, absorption is equal.

14. Power supply

It decreases the line voltage of the instrument. The converts A.C to D.C. It smooths out many ripples which may occur in the line voltage.

Figure 16.
Sample cells.

15. Types of spectrophotometers

15.1 Single-beam system

UV radiation is delineated by using the source. A convex lens accumulates the beam of emission and focal point it on the inlet splinter. The inlet splinter allows light from the source to bypass, however blocks out stray radiation. The light then reaches the monochromator, which splits it up consistent with wavelength. The exit splinter is positioned to permit mild of the required wavelength to skip thru. The chosen radiation passes through the pattern cells to the detector, which measures the depth of the radiation attaining it.

Next, to differentiate the depth of radiation preparatory to stop after it passes through the pattern, it is feasible to degree several radiation is absorbed by the pattern on the unique wavelength used. The output of the detection is commonly recorded on graph paper.

The drawback is that estimates the whole quantity of mild accomplishing particular detector, as opposed to particular proportion wrapped. The source of intensity may vary with changes in line voltage. For example, when the line voltage decreases, the intensity of the light coming from the source may decrease. The single-beam spectrophotometer is depicted in **Figure 17**.

15.2 Double-beam system

The radiation from the supply is authorised to skip thru a reflect device to the monochromator. The activity of the monochromator is to permit a slender variety of wavelengths to skip continuously an go-out slit. The radiation popping out of the monochromator through the go-out slit is received via the rotating zone which divides the beam into, one glancing through the reference and the opposite through the sample cellular. After glancing through the sample and reference mobile, the light beams are focussed onto the detector.

The yield of the detector is hooked up towards a development touchy amplifier which reciprocates to any trade-in transmission through sample and reference. The segment empathetic amplifier transmits the indicators to the recorder that is accompanied with the aid of the motion of the pen or chart. The chart drive is to integrate the rotation of the prism and for this reason, the optical density or transmission of the pattern is set down as a characteristic of wavelength.

The advantage is not necessary to continually replace the blank with the sample or to zero adjust at each wavelength. The ratio of the powers of the sample and

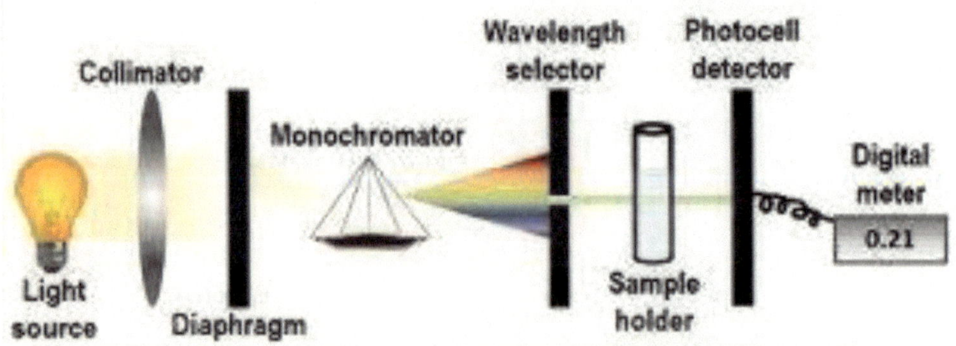

Figure 17.
Single beam spectrophotometer.

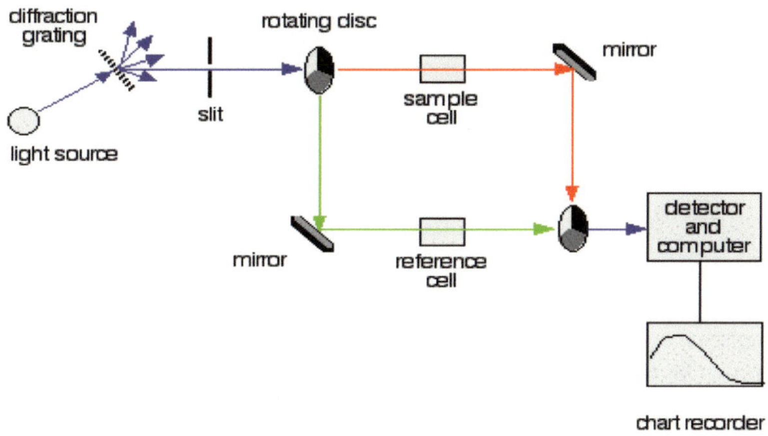

Figure 18.
Double beam spectrophotometer.

reference beams is constantly obtained and used. Any error due to variation in the intensity of the source and fluctuation in the detector is minimised (**Figure 18**).

16. Applications

16.1 Detection of conjugation

It enables to identify the relationship between the exceptional groups, especially with appreciate to conjugation can be among or extra carbon–carbon (double or triple) bonds, between carbon–carbon and carbon–oxygen double bonds and between double bonds and at an aromatic ring [11].

16.2 Detection of geometrical isomers

The trans isomers exhibit λmax at slightly longer wavelength and feature larger extinction coefficients than the Cis isomers. Examples Stilbenes in trans isomers show λmax at 294 nm, while the λmax Cis isomer has 278 nm. Detection of functional groups: to detect the presence of certain functional groups is possible, like conjugation, carbonyl group, and benzene ring.

Molecular weight determination: the molecular weight is determined. For example, the molecular weight of any amine is converted into amine picrate. Then a known concentration of amine picrate is dissolved in a litre of solution and its optical density is measured at λmax at 380 nm.

16.3 Dissociation constants of acids and bases

Consider an Acid (HA), it undergoes dissociation in water to form H_3O^+ and A^-, i.e.,

$$HA + H2O \rightarrow H_3O^+ \text{ and } A^-$$

16.4 Tautomeric equilibrium

To determine the percentage of various keto and enol forms present in a tautomeric equilibrium. Example: Ethyl acetoacetate in keto form has λmax 275 nm and ε = 16. This has only weak $n \rightarrow \pi^*$ band of the isolated carbonyl group. The enol form has λmax 244 nm and ε = 16,000, we can measure the proportions of tautomers present in ethyl acetoacetate.

16.5 Detection of impurities

The presence of impurities can be determined by the additional peaks and can be compared with that of standard raw material.

16.6 Structural elucidation

The presence or absence of unsaturation, the presence of heteroatom like S, O, N or halogen can be determined.

It is used to find out the percentage purity of samples from the formulations or raw material.

17. Conclusion

Ultra violet/visible spectroscopy is an analytical technique that is used to determine qualitatively and quantitatively for the estimation of different ions. It is a powerful technique for resolution enhancement when signal overlaps or interference occurs. This technique may also be used in many other industries. For example, measuring a colour index is useful for monitoring transformer oil as a preventative measure to ensure electric power is being delivered safely.

Author details

Sonia Karuppaiah* and Manikandan Krishnan
Department of Pharmaceutical Analysis, SRM College of Pharmacy, SRM IST, Kattankulathur, India

*Address all correspondence to: soniapharm68@gmail.com

References

[1] Skoog DA, Holler FJ, Crouch SR. Principles of Instrumental Analysis. 6th ed. Belmont, CA: Thomson Brooks/Cole; 2007. pp. 169-173

[2] Leong YS, Ker PJ, Jamaludin MZ, Nomanbhay SM, Ismail A, Abdullah F, et al. UV-vis spectroscopy: A new approach for assessing the color index of transformer insulating oil. Sensors. 2018;**18**(7):2175

[3] Ilev I, Waynant R. Ultraviolet Spectroscopy and UV Lasers. New York: Marcel Dekker; 2002

[4] Wavelength Accuracy in UV/VIS Spectrophotometry [Internet]. Available from: https://www.mt.com/ch/en/home/library/white-papers/lab-analytical-instruments/wavelength-accuracy-uvvis.html [Accessed: 5 September 2021]

[5] Metha A. Principle. PharmaXChange.info [Internet]. 2011. Available from: http://pharmaxchange.info/press/2011/12/ultraviolet-visible-uv-vis-spectroscopy-principle/ [Accessed: 8 September 2021]

[6] Atole DM, Rajput HH. Ultraviolet spectroscopy and its pharmaceutical applications—A brief review. Asian Journal of Pharmaceutical and Clinical Research. 2018;**11**(2):59-66

[7] Metha A. Derivation of Beer–Lambert Law. PharmaXChange.info [Internet]. 2012. Available from: http://pharmaxchange.info/press/2012/04/ultraviolet-visible-uv-vis-spectroscopy-—-derivation-of-beer-lambert-law/ [Accessed: 8 September 2021]

[8] Metha A. Limitations and Deviations of Beer–Lambert Law. PharmaXChange.info [Internet]. 2012. Available from: http://pharmaxchange.info/press/2012/05/ultraviolet-visible-uv-vis-spectroscopy–limitations-and-deviations-of-beer-lambert-law/ [Accessed: 8 September 2021]

[9] IEEE I, MSS, IEEE SS. In: IEEE Instrumentation and Measurement Technology Conference; New York, USA. 1994. p. 1000

[10] Reserved M-TII All Rights. Spectrophotometry Applications and Fundamentals. www.mt.com [Internet]. Available from: https://www.mt.com/us/en/home/library/guides/laboratory-division/1/uvvis-spectrophotometry-guide-applications-fundamentals.html [Accessed: 8 September 2021]

[11] Atole D, Rajput H. Ultraviolet spectroscopy and its pharmaceutical applications—A brief review. Asian Journal of Pharmaceutical and Clinical Research. 2018;**11**:59

[12] Chatwal GR. Instrumental methods of chemical analysis. Pg.no. 2.116-2.122

[13] Sharma YR. Elementary Organic analysis, Principles and chemical applications. Pg.no. 12-14

Chapter 3

Reflectance Spectra Analysis Algorithms for the Characterization of Deposits and Condensed Traces on Surfaces

Ran Aharoni, Asaf Zuck, David Peri and Shai Kendler

Abstract

Identification of particulate matter and liquid spills contaminations is essential for many applications, such as forensics, agriculture, security, and environmental protection. For example, toxic industrial compounds deposition in the form of aerosols, or other residual contaminations, pose a secondary, long-lasting health concern due to resuspension and secondary evaporation. This chapter explores several approaches for employing diffuse reflectance spectroscopy in the mid-IR and SWIR to identify particles and films of materials in field conditions. Since the behavior of thin films and particles is more complex compared to absorption spectroscopy of pure compounds, due to the interactions with background materials, the use of physical models combined with statistically-based algorithms for material classification, provides a reliable and practical solution and will be presented.

Keywords: diffuse scattering, remote sensing, spectroscopy, surfaces, detection, mid wave infra-red, short wave infra-red, classification

1. Introduction

Spectroscopy is one of the foremost and main methods of characterizing materials of various states of matter—gas, liquid, solid, vapor, and aerosol. The need to remotely detect and identify residues, traces, contaminations, and small amounts of chemicals plays an important role in many fields, such as forensics and security [1], agriculture [2], food quality pharmaceuticals industry, climate research, and others.

Detection of surface contaminations and residues in a standoff manner enables scanning surface from a safe distance and with no physical contact with the sample, which may be hazardous or too sparse to necessitate wiping large area. Therefore, optical sensing can provide an immediate result, and it is favored over surface sampling techniques such as mass spectrometry techniques. Several studies and efforts utilizing probing techniques for ion-mobility spectrometry of explosives form surfaces through surface wiping [3] plasma ionization [4], desorption electrospray ionization from skin [5], and more. Non-contact preconcentration and ionization sampling methods, such as airflow-assisted ionization [6] and laser-induced breakdown spectroscopy [7], enable standoff detection of surface absorbed chemicals. For further reading on sampling techniques and explosive detection, refer to Tourné's review [8].

Fourier-transform infrared (FTIR) spectroscopy, coupled with attenuated total reflection (ATR), can be used to probe trace amounts of spores [9] or other particles on solid surfaces [10], or remotely by imaging [11, 12]. Active methods for powders detection, in which the investigated sample is artificially illuminated rather than using ambient light, were also applied in various spectral regions, such as THz [13, 14], but more commonly in lower wavelengths. The mid-IR region, the most common for chemical spectroscopy, was pushed forward by developing the quantum cascade laser (QCL) [15, 16]. Laser-assisted spectroscopy for explosive detection, and other particulated matters, was also employed by various optical techniques, such as photo-acoustics [1, 17, 18]. Using near-field optical microscopy, Craig et al. scanned surfaces with condensed residues by scanning QCL with a rapid acquisition time of 90 s per spectrum [19]. Explosive's particles and residues using scanning QCL microscopic hyperspectral imaging enabled a four-second acquisition time [20]. Advancing technology provides the means for an even shorter detection time in the range of <0.1 s per spectral cube of trace surface contaminants [21]. A similar method, i.e., imaging the diffusely scattered radiation from a light source, was also demonstrated using a CO_2 laser [22]. The short wavelength of the IR region was also used for screening envelopes for traces of hazardous powders by hyperspectral imaging [23]. Multispectral imaging in the visible range was demonstrated for the detection and discrimination of bloodstains on cotton [24].

IR spectroscopy in the 7–14 μm band, i.e., mid-wave IR (MWIR, also termed long-wave IR; LWIR), is discussed in the first section of this chapter. This spectral region is used for inspecting molecular chemical information for gaseous and condensed matter. In condensed samples, MWIR spectroscopy is an analytical technique fitted for trace analysis, but it is non-penetrating, which can be affected by the surface morphology that scatters the light and affects the resulted spectrum. Short IR wavelengths (SWIR, 1–2.5 μm) spectroscopy, addressed in the second section of this chapter, is characterized by weaker absorption and larger penetration depth, resulting in diffuse reflection, allowing deeper investigation of the sample and revealing its physical and chemical structure. This is not an analytical method but provides information not only on the sample's surface but also on the sample's core compound and structure.

This chapter is divided into two main sections. The first section elaborates the issues of MWIR spectroscopy of surfaces and suggests a practical solution based on a physical model. The second part confronts the challenges of using SWIR for similar purposes and proposes a solution based on statistical learning.

2. Reflection and scattering: MWIR

The MWIR region, known as the 'molecular fingerprint region', reflects the molecular composition of most chemicals vib-rotational transitions, hence reviling the molecular structure by its unique spectral fingerprint. Due to its analytical capabilities, it was favored by spectroscopists for the investigation of the chemical properties of organic materials, and detection and identification needs. Acquiring spectrum for gaseous samples is quite straightforward—launching an optical beam through the sample results in spectral attenuation, i.e., absorbance, caused by the molecular transitions. The absence of back reflection is due to the lack of specific boundaries between materials with different refractive indices (n_1 and n_2). However, in a condensed matter, where the boundaries between materials are well defined such as in layers and particles, interference scattering and diffraction effect the spectral scattering. Each of the following phenomena: interference, diffraction, and scattering, has its theory models and approximations that describe an interaction of light and matter. In fact, these three phenomena are manifestations of maxwell's equations, differ only by mathematical approximations, and therefore provide ambiguous results at certain conditions.

The simplest case is the reflection from a uniform layer, explained and demonstrated in the following sub-section, which has an exact solution and can be expanded to cases of non-uniform layers, such as traces and residues, as depicted in **Figure 1**. The upper figure illustrates a light beam that is specularly reflected from a flat layer. The lower figure presents a surface with residual contamination, which results in diffuse reflection. The lower figure presents the illumination of a contaminated surface. The surface is tilted such as the specular reflection component (in red), which can be orders of magnitude stronger than the diffuse reflection component, is directed away from the detector. Part of the diffuse reflection lobe, which originates mainly from residual traces, is directed to the detector (blue). This section explores the physics of MWIR reflection from a uniform and non-uniform coverage of surfaces and suggests methods for detecting and identifying traces and residues.

2.1 Uniform layers: general

As mentioned, the reflection of light from surfaces can be generally divided into specular reflection and diffuse reflection components, both illustrated in **Figure 1**.

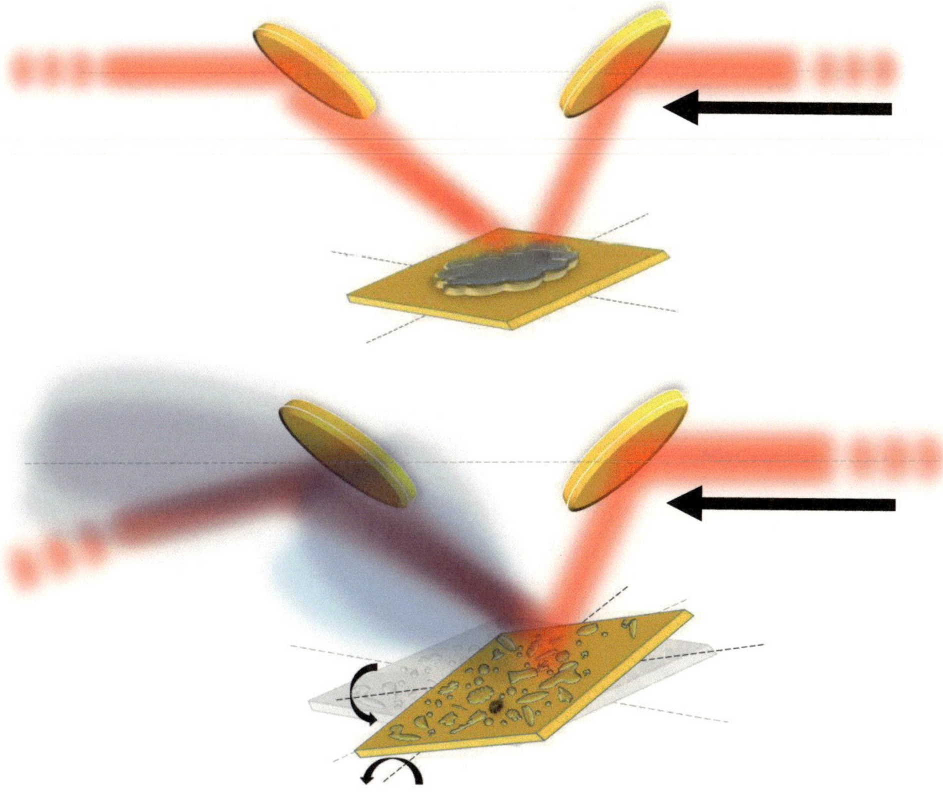

Figure 1.
Reflected light from residuals on a surface. The upper panel illustrates a uniform layer deposited on a surface and its specular reflection. The lower panel illustrates drops, particles and residuals on a surface and the diffuse reflection measurement. The black arrows represent the incident light direction.

The upper panel of the figure presents specular reflection from a finite layer, where the reflected light is directed to a detector. In thick and highly absorbing layers, the measured spectral reflection is comprised exclusively by the layer's molecular properties, i.e., the absorption coefficient and refractive index as described by Fresnel equation [25]. When the absorption coefficient and the thickness (and the light coherence length) allow the substantial optical intensity to be back-reflected from the carrying surface and complete at least one round trip, interference effects are observed, as illustrated in **Figure 2**. The figure illustrates the tracing of an optical field hitting a finite smooth layer with incidence angle θ. The multiple reflections and refractions amplitudes are summed up and then squared to get the light intensity. Almost no surface is smooth enough to avoid the scattered diffuse reflection lobe (illustrated in **Figure 1** lower panel). This component, which obeys the Lambertian reflection law with highly rough surfaces, is much weaker and negligible for relatively smooth surfaces. In cases of a partial cover of a surface by an analyte, it is desired to avoid the specular reflection, which might be orders of magnitude stronger than the diffuse reflection of the analyte traces and therefore mask it, as illustrated in **Figure 1b** and discussed in Section 2.3.

Figure 2 refers to the scenario illustrated in **Figure 1a**, and presents detailed optical ray tracing through a finite and uniform thickness layer. An optical beam is hitting a uniform interface of a transparent layer with parallel facets at incidence angle Θ. Inside the layer, the optical field suffers multiple internal reflections, which give rise to interference. The reflection of a non-absorbing layer is:

$$R = \frac{\left(\sqrt{R_1} - \sqrt{R_2}\right)^2 + 4\sqrt{R_1}\sqrt{R_2}\sin^2(\delta/2)}{\left(1 - \sqrt{R_1}\sqrt{R_2}\right)^2 + 4\sqrt{R_1}\sqrt{R_2}\sin^2(\delta/2)} \tag{1}$$

where R_1 and R_2 are the intensity reflection coefficient from the facets (calculated by Fresnel relations) and the optical phase is $\delta = 2\pi n_{(\lambda)} L \cos(\Theta)/\lambda$ for a layer with thickness L. It is clear that unlike absorption spectroscopy, which measures the attenuation directly, the interference described in Eq. (1) has a crucial impact on the reflected spectra. Generalizing Eq. (1) to the case of semi-absorbing material, for example, a nonvolatile liquid over a glass window, ceramic tile, metallic surface, or other casual flat surfaces, results in the following equation:

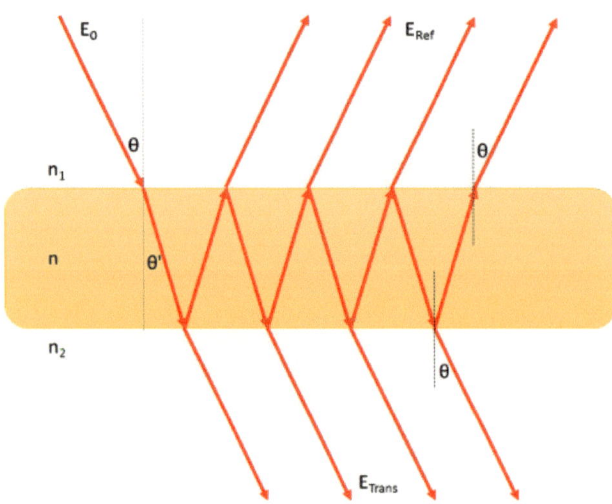

Figure 2.
Ray tracing of a finite uniform layer. E_o, E_{Ref}, E_{Trans} are the incident, reflected, and transmitted fields, respectively.

$$R = \frac{\left(\sqrt{R_1} - \sqrt{R_2}e^{\alpha L\cos(\theta)}\right)^2 + 4\sqrt{R_1}\sqrt{R_2}e^{\alpha L\cos(\theta)}\sin^2(\delta/2)}{\left(1 - \sqrt{R_1}\sqrt{R_2}e^{\alpha L\cos(\theta)}\right)^2 + 4\sqrt{R_1}\sqrt{R_2}e^{\alpha L\cos(\theta)}\sin^2(\delta/2)} \tag{2}$$

where α is the attenuation coefficient defined as $4\pi k_{(\lambda)}/\lambda$ (λ is the wavelength, and $k_{(\lambda)}$ is the imaginary part of the refractive index), thus, the reflected spectrum is affected heavily by the layer thickness, as seen in **Figure 3**, showing the absorption of Polymethyl Methacrylate (PMMA) layer, and reflected spectrum from different layers of it. The inset of the figure shows the absorption coefficient, measured with ATR. The figure shows reflectance measurements of three different spin-coated layers of PMMA, measured with a spectrally scanning laser, according to the set-up depicted in **Figure 1**. The measured spectra cannot be precisely associated with the absorption coefficient, from the reasons described above (Eq. (2)). Significant differences between the three layers can be seen as the peak's location are shifts and change their shape. This figure accentuates the resulted differences of reflected spectra from the same material with different morphology (i.e., layer thickness). The thickness differences are just a few microns (layers thicknesses are 16, 20, 27 μm), indicated by low correlations of the measured signatures, as shown in the table given in the below figure.

2.2 Uniform layers: experimental implementation

As explained above, by knowing the optical properties of the layer and the carrying surface, one can calculate the reflected spectrum, using Eq. (2). This is exemplified in **Figure 4**, showing the real and imaginary parts of the refractive index of poly-dimethyl-siloxane (PDMS) on a metallic surface (**Figure 4a**), and the measured spectral reflection from 0.63 μm to 22 μm layers (**Figure 4b** and **c**). **Figure 4a** presents the measured ATR spectra of PDMS, where the refractive index (n) is calculated by Kramers-Kronig relations [26]. **Figure 4b** presents the reflected

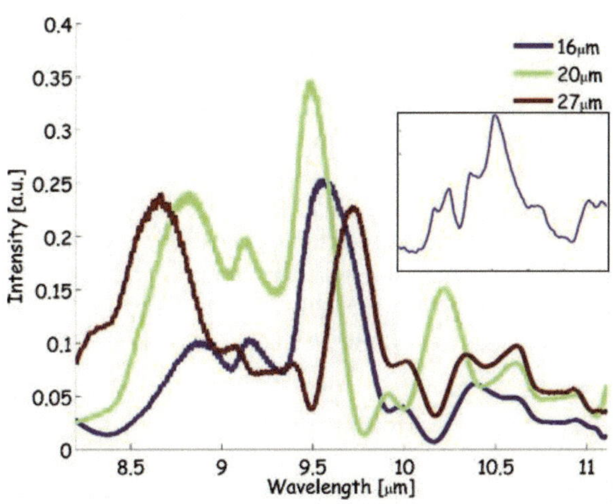

Corr (16 μm , 20 μm)	Corr (16 μm , 27 μm)	Corr (20 μm , 27 μm)
0.1	0.25	0.63

Figure 3.
Reflected spectra of PMMA layers. The inset depicts the absorption coefficient, and the three curves of the figure represent the reflected intensity from three different uniform layers. The table represents the correlation coefficients between the layers.

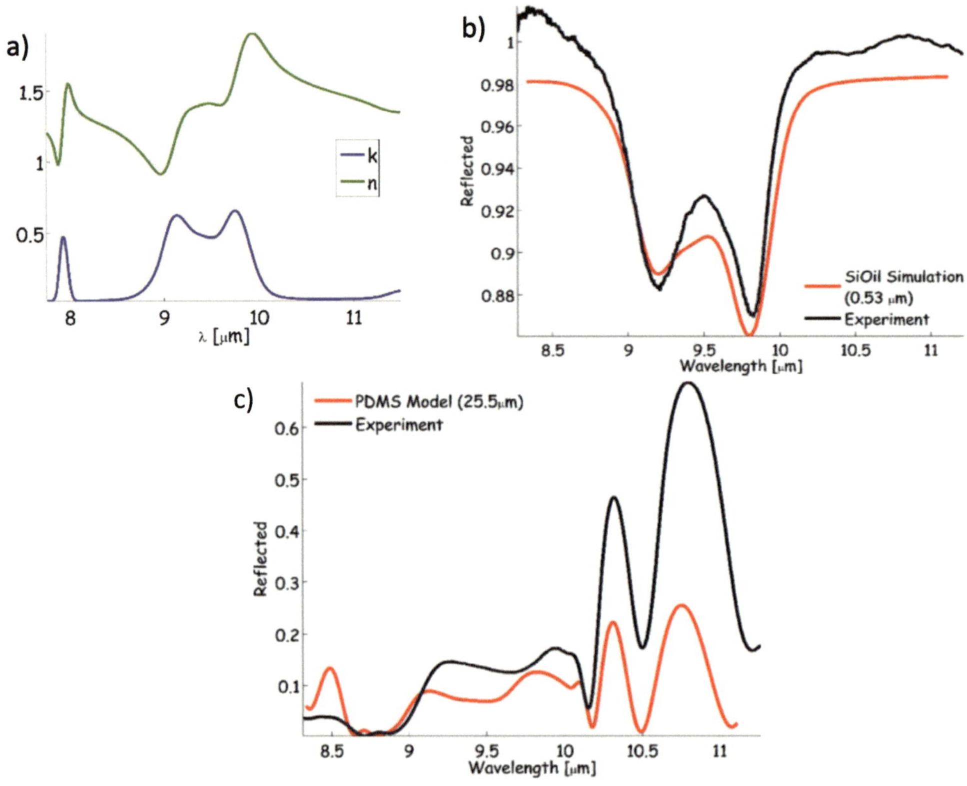

Figure 4.
PDMS layers reflection spectrum. (a) Optical properties (n is calculated). (b) Thin layer. (c) Thick layer.

spectrum of a thin layer that matches the absorption coefficient 'k' from **Figure 4a**. This layer is too thin compared to the wavelength, therefore interference effects are not observed, and the reflected spectrum correlates well with the absorption coefficient. **Figure 4c** presents the reflected spectrum from a layer having thickness more than twice the average wavelength. Consequently, an interference pattern appears at the edges of the acquired spectrum, where the absorption is negligible. Also, some deformation of the absorption around 9.5 μm is noticed. The interference fringes properties can be used for exact evaluation of the layer thickness by Eqs. (1) or (2)—the spacing between adjacent peaks (called 'free spectral range') is affected by the wavelength, thickness, incidence angle, and the refractive index, which are the components of the phase parameter δ from Eqs. (1) and (2). For a more accurate solution, one should account for the variation in the Snell law for absorbing medium interface, which may affect the form of Eq. (2) [27].

2.3 Non-uniform layers: diffraction and scattering

Disseminating liquid materials more efficiently can be achieved by spraying and drizzling, which cover larger areas with drops or droplets. Such dispersal processes usually result in size distribution similar to lognormal [28], which can be very wide and with standard deviation spread from a few microns to hundreds of microns. The size distribution of sprayed droplets has an important effect on the performance of agrochemical systems [29] and combustion engines [30] etc., mainly characterized by scattering and diffraction measurements of levitating particles. Mostly, the efforts are towards covering large volumes and surfaces with droplets for the highest efficiency of dissemination and surface coverage. The results are similar to the lower panel of **Figure 1**, illustrating residuals and traces on a planar surface. The surface is tilted to avoid the specular reflection component that might mask the diffuse reflection component, which can be orders of magnitude weaker.

Light scattering is an extensive and well understood physical phenomenon, originating from Maxwell equations in the form of the wave equation [31], and depends on many factors such as the wavelength of light, incidence angle, material properties of the scatterer (absorption and refraction), and geometric factors of scatterers and illumination. As such, many scattering theories and models were developed, each describes these phenomena under different conditions and assumptions. For example: (1) Mie theory (full name: Lorenz-Mie-Debye theory) describes general scattering by homogeneous, isotropic spheres with no size limits but is more commonly used where the scatterer size is comparable to the light wavelength [32], and have many further approximations for different sizes and shapes of scattering particles. (2) Rayleigh scattering theory (Rayleigh-Gans-Debye) describes scattering from particles smaller than the wavelength [33], and more. In cases where the dissemination process, and hence the resulting size distribution, is unknown, it is unclear how to choose the most suitable approximation. Moreover, in liquid spray sediment over a surface, the interaction between the surface and the droplets may dramatically change the size and geometry of the droplets, and each scattering model might result in a different solution. We should note that most of the interactions of light with matter are fundamentally the same, and all are described by Maxwell's equations. More specifically, interference is a basic outcome of these equations, and it is the cause of diffraction and scattering, which are all different manifestations of light interaction with matter, and the only difference is the approximation of different theories. Therefore, it is expected that scattering by a sphere and reflection by a slab are similar [34, 35]. Accordingly, it will be demonstrated that the above-presented layers model (LM) can be used in many realistic scenarios for the detection and identification of sprayed liquid on a surface. Similarly, it is suggested that unknown condensed residuals on a surface can be detected, identified, and quantified by a simple reflection model instead of a complicated specific scattering model.

Figure 5 presents measured and calculated diffuse reflection normalized spectra of PDMS, the solid black spectra were calculated using the LM presented in the previous section, and the red dashed curves represent measure spectra of laboratory

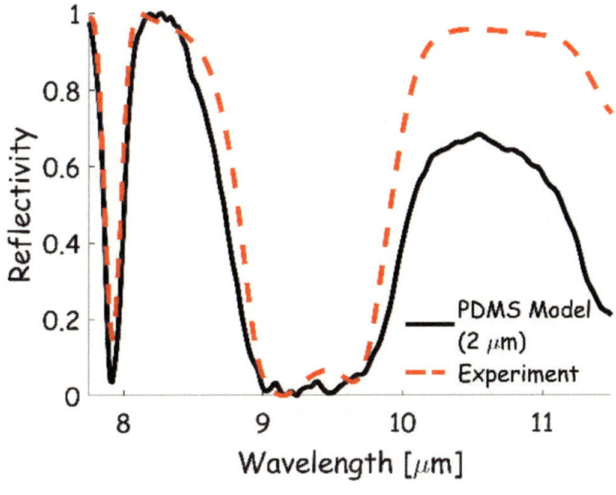

Figure 5.
Diffuse reflection measurement (solid black) of PDMS spray-on metalic surface, vs. LM (dashed red) (from [36]).

experiment, where PDMS was sprayed on a metallic surface and measured with EC-QCL, using the apparatus of in **Figure 1**. The droplet's size distribution was between a few microns and dozens of microns, where the majority was 10 microns and less. To resolve the thickness parameter (L) for the LM (see Eq. (2), where $\delta = \delta(\lambda)$), the LM was calculated repeatedly for all the desired thickness range and correlated to the experimental results, while the best correlation is presented in the figure. Non-uniform coverage means non-uniform thickness and non-continuous coverage; hence the LM can provide an estimation of the 'effective thickness' (ET) and relative surface coverage using the following:

$$I(\lambda)_{total} = \alpha I(\lambda)_{PDMS} + (1 - \alpha) I(\lambda)_{Subs} \quad (3)$$

where α is the coverage factor, I_{total} I_{PDMS} and I_{Subs} are the total spectral measurement, PDMS contribution, and reflected intensity of the substrate, correspondingly. It was found that the lab experiment corresponds to 2 μm ET and 2% coverage.

As demonstrated in **Figures 3** and **4** for layers, and in **Figure 5** and Ref. [36] for sprayed traces, the spectral scattering of different ET expresses in non-identical spectral responses, i.e., different peaks heights and locations. **Figure 5** and Ref. [36] exemplify that fitting the best correlated LM result to unknown dissemination results in high certainty identification [36]. Using this approach of forcing the LM over non-uniform coverage, we can circumvent the requirement of the different scattering models to characterize the morphological properties of the analyte and provide identification and also a good estimation of its coverage features.

3. Detection and identification in the SWIR

Although, the absorption spectrum of the invisible range was discovered in 1800 was at the near-infrared (NIR) [37–39], it was almost ignored until the 1950s, as it was not considered for analytical purposes. Spectroscopic investigation of chemicals, for various purposes, was perused in the longer wavelengths, i.e., 7–14 μm, which provides specific information regarding the molecular structure of materials, which are expressed in distinctive narrow absorption peaks. The spectroscopy in the VIS-SWIR range (sometimes referred to as NIR, 800–2500 nm) exhibits broad overlapping peaks resulting from combinations of overtones and transitions in the MWIR. Theoretically, under the assumption of harmonic potential, part of the transitions is forbidden, resulting in low absorption. In the 1950s, SWIR spectroscopy was pushed forward due to technical and analytics progress, i.e., the development of lead sulfide semiconductor detectors and improvement in computing and data analysis techniques—chemometrics [40]. Consequently, this spectral range became more prominent in industry and research, providing rapid, nondestructive measurements for various fields [41, 42]. Another difference between SWIR and MWIR is the increased penetration depth of SWIR which effects the diffusion of the reflected light, as illustrated in **Figure 6**. As seen from the figure, the reflection of light from a solid sample occurs in two forms—specular and diffuse reflection. In contrast to MWIR, which is reflected from the surface, radiation in the SWIR will penetrate the sample and therefore is affected both by absorption and scattering. Various studies utilized this property for the investigation of deeper layers of samples, such as fruits and vegetables having penetration depth between 1 and dozens of microns [43, 44], or near-surface (~1 mm) as in soil [45]. Another advantage of using SWIR is the lack of blackbody radiation from the sample and the high SNR of detectors. For example, among other advantages and

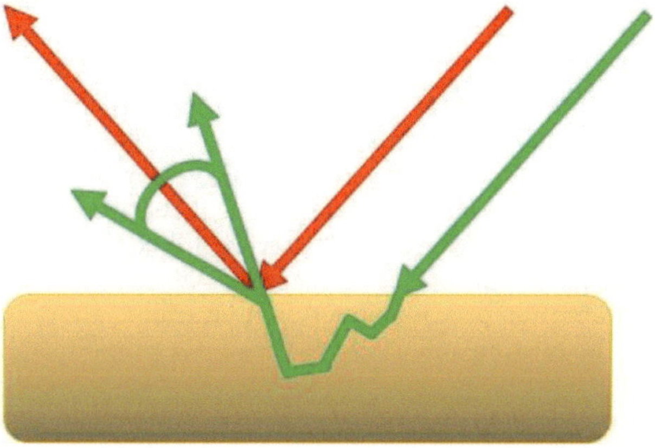

Figure 6.
Reflection measurements—spectral (red) and diffuse (green) reflection. In cases where the beam penetrated the sample, it suffers.

uses of SWIR spectroscopy, Manely indicated the increased penetration depth (compared to the MWIR), that can be utilized to rapidly analyze biological materials such as food products without sample preparation [41].

Reflectance measurements produce a large amount of data that have to be analyzed in real-time. For example, Park et al. utilized reflectance spectroscopy for real-time in-line poultry fecal detection [46]. They placed a dedicated camera to measure the reflectance spectrum in every pixel in the image—a hyperspectral imager (HSI—see **Figure 7**). The HSI was integrated into the production line, so both the data acquisition and interpretation are performed in real-time with no need for exhaustive sample preparation techniques. Similarly, Sendin et al. described the application of SWIR HSI for the quality and safety evaluation of cereals which are an essential part of the global population diet [42]. The HSI measurements can be performed in various scales enabling monitoring of a single grain to detect, for example, fungi, or on a larger scale to determine the overall cereal stock quality. Another area in which spectroscopy in the SWIR gained popularity is geophysical mapping. Goetz, one of the pioneers in this vibrant field of

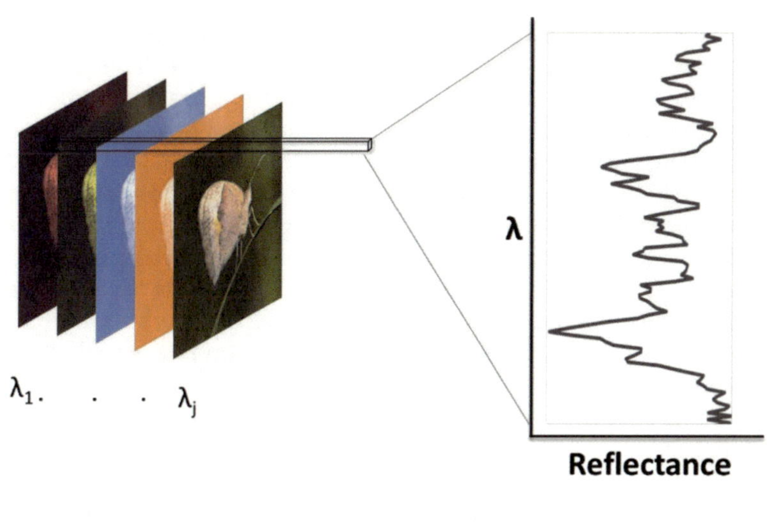

Figure 7.
An illustration of the HSI measurement. A scene is sampled in various wavelengths resulting in multiple images, which are referred to as a data cube. Each pixel in the data cube contains both spatial and spectral information. The right side of the figure shows the reflectance spectrum extracted from a pixel in the upper left corner of the image.

research, recognized the future advantages of imaging remote sensing in the 1980s, for the identification of earth surface materials by their reflectance spectra [47]. For example, Ben-Dor et al. mapped soil properties using air-born HSI in the visible SWIR range (DAIS-7915) [48]. They developed methods to map soil organic matter, soil field moisture, saturated moisture, and soil salinity. These studies and many others utilize the robustness of the spectral measurements in the SWIR. However, two challenges have to be considered; the first one is the atmosphere effect of the reflected light, and the second one is data analysis. Brook and Ben-Dor, developed a calibration technique that utilizes targets placed in the trajectory of the airborne sensor [49]. The calibration process (supervised vicarious calibration—SVC), is performed during the mission and accounts for sensor properties and atmospheric interference.

Data analysis also received significant attention, and several algorithms have been introduced over the years to detect sample anomalies or specific target materials in imaging and non-imaging spectroscopy. Manolakis and Shaw described several algorithms for analyzing HSI data [50]. In general, these algorithms are designed to find indications for a phenomenon of interest. For example, the presence of a specific material that exceeds the naturally occurring variability in the sample, which is to be expected in many real-life applications such as process control, geophysical mapping and other applications in which sample preparation is impractical. Some of these algorithms can resolve common situations in remote sensing, where the target material occupies only a portion of the sampling area, i.e., linear mixing between the reflectance spectrum of the target material and unknown background material. However, in some specific cases, non-linear mixing may also occur since the incoming light might pass through a thin film of one material and then be reflected by the background material that supports this film—"intimate mixing", in which the resulted spectrum is a dot product of the film and the supporting material [51]. Kendler et al. developed an algorithm that automatically resolves non-linear mixing between background and target materials by utilizing the benefits of HSI [23, 52]. The algorithm seeks a pixel without the target material (clear background) having a similar reflectance spectrum supporting the target material (target pixel). Once such clear background is located in the image, the pure spectrum is extracted and compared to a library reference. It was also shown that the quality of extraction of the pure spectrum increases as the physical distance between the clear background and the target pixel decreases [53]. Conversely, in non-imaging spectroscopy, such a process may be impractical. This is exemplified in **Figure 8**, presenting normalized laboratory measurements of commercial granulated sugar and powdered sugar (<125 μm), disseminated over PVC and laminated wood surfaces. The measurements were conducted with a spectrometer (ASD FieldSpec® 4 Hi-Res) at 350–2500 nm using a custom accessory sampling contact probe consisting of a Halogen lamp and a collecting fiber as input. This tabulated figure accentuates the diversity and complexness of SWIR reflectance spectroscopy described above, by presenting significant spectral differences of four different types of measurements of the same chemical. It shows that both the surface and the physical state of the sample affects the reflected spectrum in a way that cannot be eliminated using a simple non-linear unmixing model.

Therefore, it is concluded that utilizing SWIR spectroscopy for a minute amount of material placed on a surface might pose a significant analytical challenge. Although, a simple unmixing model can be applied in some cases, where the background surface is opaque, the sample has low light scattering and absorption, and the measurement is performed using an imaging spectrometer. However, it is hard to guarantee such a set of conditions in many cases that may be ubiquitous in real-life applications.

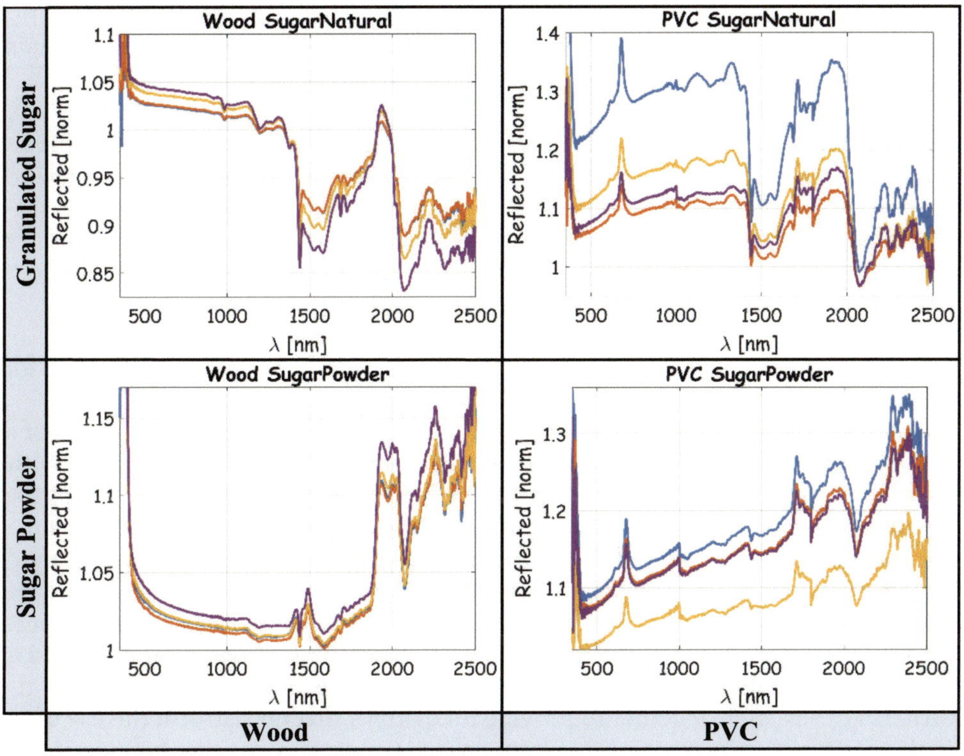

Figure 8.
Diffuse reflectance of sugar (granulated and powdered) on PVC and wood surfaces.

3.1 Classification of powdery residuals

Spectral detection and classification of powdery and condensed materials for security, safety, food industry hygiene, etc., was suggested at various spectral regions, from THz [13, 14, 54, 55] through the MWIR [12, 34, 35], SWIR [56, 57] and more. However, in lower wavelengths (VIS-SWIR), most powders, especially fine-grained powders, appear visually similar and therefore hard to distinguish, have little color and texture, and appear white (due to scattering). Also, the spectral contribution of the carrying surface tends to blend with the powder's spectrum, as discussed above. The literature shows many examples for powders and materials detection and classification, utilizing the availability and low cost of using the visual and near IR ranges for that task. Spectral imaging and computer vision was employed by Zhi, who showed reasonable classification accuracy (60–70%) for discrimination of 100 powders, using three cameras (RGB, NIR, SWIR) and 12 minutes acquisition time [58]. Classification of surfaces materials, also by computer vision, was used by capturing the micro-geometry and reflectance properties of the surfaces, using a photometric stereo sensor with a 3 cm working distance [59]. These examples require expensive light sources or complicated measurement apparatus, despite the use of the visual regime, which is supposed to be cheaper and less complex. This section presents the VIS-SWIR non-imaging spectral measurements and data analysis approach, for powders on various surfaces, with low hardware and software requirements, for powders detection and identification through classification.

3.1.1 Data acquisition and analysis

Different powders (sodium carbonate, Arizona dust, tryptophan, tenax, caffeine, hepes, copper) were disseminated on various surfaces (ceramics, laminated wood, PCV, cardboard, Perspex, acrylic glass, Teflon, painted and bare car metal, plastics, marble stone, pebble), and their VIS-SWIR spectrum was acquired using a non-imaging spectrometer. The data were collected using a spectrometer (Filedspec4 from ASD, with 2151 sampling points from 350 to 2500 nm) and a sampling contact probe consisting of a Halogen lamp and a collecting fiber as input. For each surface, various measurements were conducted, having different concentrations of powder per unit surface. Thus, a large dataset was collected, comprised of a range of intensities, and different relations between absorption peaks. Another source of spectral diversity is the size, shape, and orientation of the measured particles, which affects their scattering properties. It is worth noticing that due to these scattering effects, the orientation (pivot) of the surface also affects its reflected spectrum. These measurements produce a diverse dataset, well suitable for statistical learning methods, aka machine learning.

Supervised learning (classification) is the process of predicting the category of a given data point, based on a label train-set. The learning algorithm is supplied with a labeled train-set, in which every measurement is labeled in advance and learns the mapping function between the new input data and its response (label). Some data sets are linearly separable, and can be processed with linear algorithms such as SVM (support vector machine), while other problems require a higher-order hyper-plane to resolve the data. Applying some effort (such as kernel methods) enables linear learning algorithms to learn non-linear decision boundaries. Another approach to 'upgrade' simple algorithm performance is ensemble learning, in which weak learners are ensemble together to provide predictions that outperform the use of a single type of these learners, and to learn more complicated decision boundaries. One of these algorithms is random forest (RF, also known as tree bagger), which uses classification trees [60] and bootstrap aggregation random processes. The RF algorithm is trained using ~70% of the data (training set), and the produced model is tested on the rest of the unseen data (test-set). On the train set data, RF performs a process of bootstrap aggregation (bagging) in which a subset of the train set (2/3) is randomly picked iteratively to produce a classification model whose results are examined on the rest third of the data- the validation set. This random process, and the random selection of predictors at each tree node, decrease the variance errors and enable the production of another measurement of the predictor—the "out-of-bag error" (OOBerr). By comparing the errors produced with and without each predictor, the OOBerr scores each predictor to note its contribution to the learning process. For further reading about RF, see Breiman [61–62].

Figure 9 presents the confusion matrix[1] presenting the classification results of the aforementioned powders and surfaces, showing total accuracy of almost 90%, and high true positive rates (TPR, noted in the right vertical column). Note that the presented powders appeared white and indistinctive to the human eye, except copper that is in the form of metallic reddish flattened flakes. Nonetheless, all powders (except tryptophan) were classified with a similar TPR. The measurements are not normalized by the surface signature, exemplifying the strength of the classification process, which learns to ignore the distracting influence of the carrying surface.

As explained, the OOBerr parameter evaluates the contribution of each predictor (i.e., wavelength) to the learning process. By using only some of the top influencers, it is possible to avoid bad predictors, and reduce the required data volume, thus

---

[1] The confusion matrix compares the true class with the predicted class for each class. The diagonal terms represent the true classification, and its ratio to the total matrix represents the accuracy.

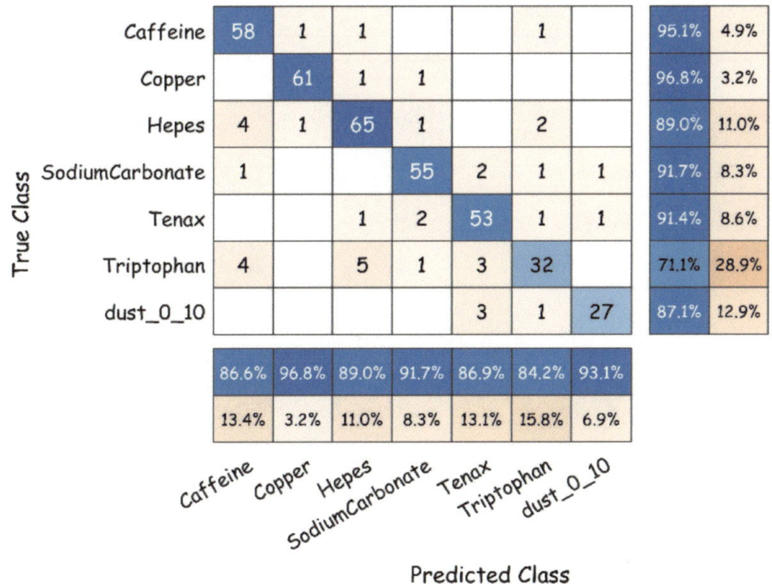

Figure 9.
Confusion matrix—classification of powders on surfaces, using hyper-spectral data.

speeding the whole process. **Figure 10b** illustrates the importance of all 2151 predictors from the OOBerr computation, where the top 5% are marked in black. The confusion matrix in **Figure 10b** shows RF classification with these 108 selected features. We see that the classification performance is almost unchanged, and the accuracy has an insignificant decrease of 0.4%.

3.2 Application: classification of wheat yellow rust disease

Wheat is one of the world's major crops and provides a substantial amount of starch, proteins, vitamins, and dietary calories worldwide. One of the major threats, on a global scale, to wheat production is Yellow Rust (YR) disease, which is the most damaging disease of wheat on a global scale and causes an annual loss of millions of tons of wheat harvest valued at around 1G USD. Due to global warming and the evolution of YR strains adapted to higher temperatures, YR damages wheat crops in areas where it had not been previously reported. The spores are carried with the winds and reaches high altitudes so it can travel long distances. YR has a complex life cycle that includes several hosts and spore stages, which eventually appear as yellow particles (size of a few dozens of microns) covering the leaf surface, as seen in **Figure 11** left. Managing yellow rust can be utilized through the application of fungicides [63, 64], using resistant plants [65], and tracking [66]. While using resistant varieties is an efficient strategy, it poses some challenges for YR detection for two main reasons: (1) co-evolution of the host wheat and the fungal pathogen, might enable YR to overcome the YR-resistant genes mechanism. Hence, resistant plants also need monitoring for disease detection. (2) The Hypersensitive Response (HR) of a resistant leaf appears to be visually similar to YR disease. This is exemplified in **Figure 11** (right), presenting HR and YR early stages, which are visually hard to discriminate. This situation resembles the above-described detection of powder on a carrying surface, where the surface is the green leaf, and the powder is the powdery particle of YR spores. Therefore, applying similar measurements and data processing methods can successfully classify the disease stages and HR response for the early detection of YR in the field.

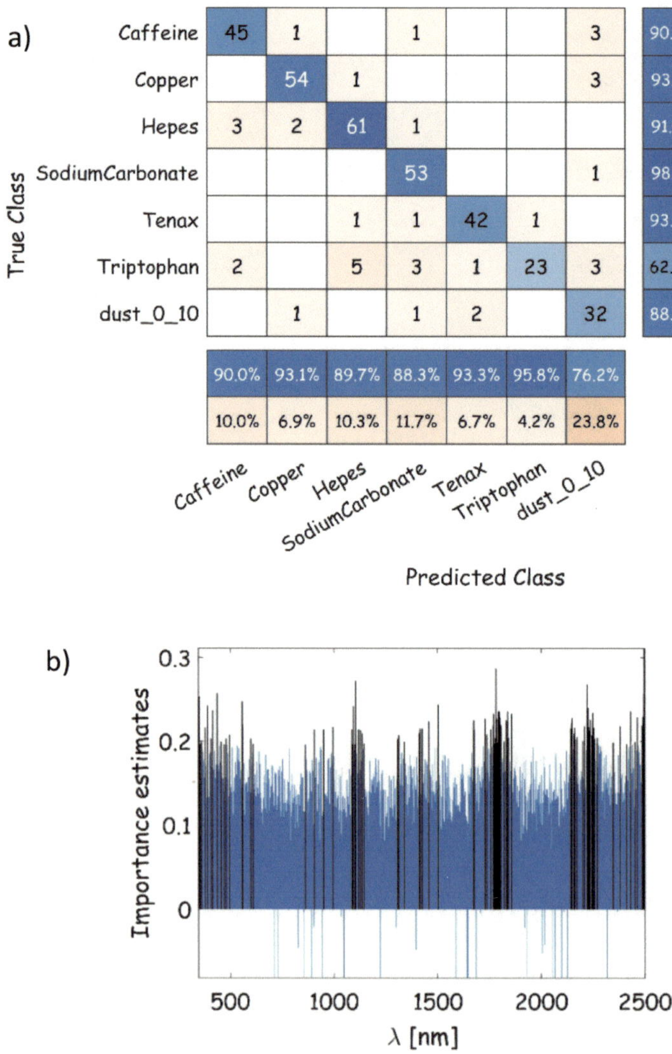

Figure 10.
(a) Confusion matrix—classification of powders on surfaces, using top 5% important predictors presented in (b).

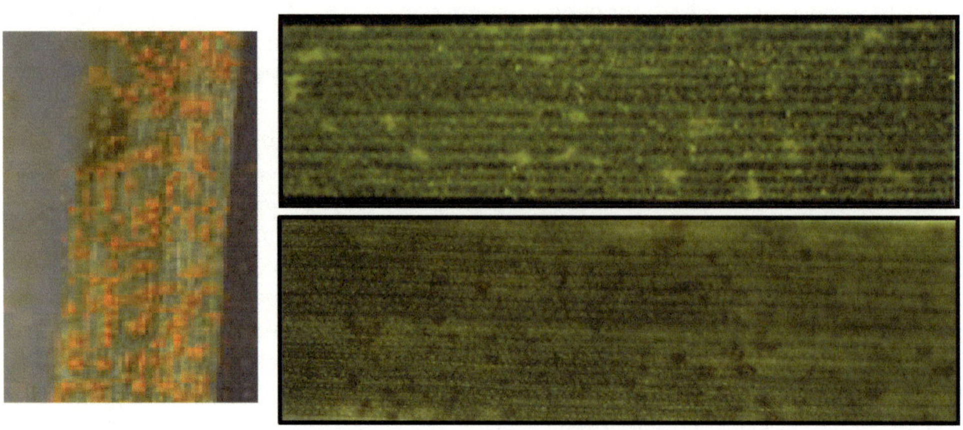

Figure 11.
YR symptoms on susceptible and resistant wheat leaves. Left—fully developed YR. Right: resistant (top) and susceptible (bottom) wheat leaves response to YR. The top leaf exhibits HR symptoms, which is visually similar to YR early symptoms in the lower leaf.

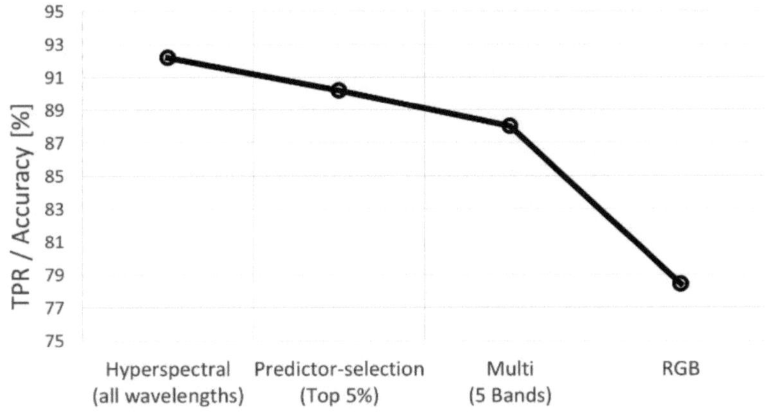

Figure 12.
True positive rates (TPR) of early stages of sporulation vs. four levels of the predictors set: hyperspectral (all 2151 wavelengths), predictor-selection (top 5% important); multi-spectral (5-bands), and RGB.

Figure 12 illustrates the results of the classification process. The VIS-SWIR spectrum of hundreds of wheat leaves in various stages was acquired and classified with RF. Some of the leaves were green and healthy, some susceptible to YR, exhibiting sporulation in several stages (from the onset of sporulation; sporulation early stages; final sporulation stages), and some resistant leaves exhibiting several stages of HR (early stages and fully developed HR). The figure presents the TPR of sporulation early stages at four different data dimensions. Using all 2151 predictors produces a TPR of 92.2%. Using the feature selection process described in the previous section produced a slight decrease (TPR of 90.2%), using only 108 predictors. Further directionality reduction to five spectral bands (conventional agricultural imager) results in a TPR of 88%. Using only RGB channels results in a TPR of 78.4%. This noteworthy result pledges that YR detection does not require expensive specific hardware, and enables on-site monitoring by non-experts.

4. Summary/conclusions

Spectral light reflection can be used for the identification of particulate and condensed chemicals. In controlled situations, such as laboratory measurements, the sample can be manipulated to provide consistent high-quality reflectance spectra that can be used to characterize and identify the sample at hand. Such manipulations involve classical sample treatment techniques such as purification, grinding, and pressing to a pellet. However, standoff sensing of un-manipulated samples results in noisy measurements, requiring more sophisticated data analysis to extract meaningful information. LWIR spectrum does contain specific, unambiguous information of the molecular structure, but it involves scattering phenomena that require an adapted model, accounting for the sample's physical geometrical properties. A simple unified model was suggested to bypass this issue using a simplified model. A different analysis method was demonstrated in the case of light scattering in the SWIR. In this case, the reflectance spectra are broader and weaker, and a machine-learning model is used to classify the sample according to its typical reflectance. An additional consequence of the machine-learning model is the assessment of the contribution of each wavelength to the accuracy of the classification. Using only the important wavelength can speed up the computation and simplify the measurement, thus enhancing the usability of reflectance spectroscopy.

List of abbreviations

ATR	attenuated total reflection
EC-QCL	external-cavity quantum cascade laser
ET	effective thickness
FTIR	Fourier transform infra-red (spectroscopy)
HIS	hyperspectral imager
HR	hypersensitive response
IR	infra-red
LM	layers model
NIRMWIR/SWIR	near/mid/short wave infrared
OOBerr	out-of-bag error
PDMS	poly-dimethyl-siloxane
PMMA	polymethyl methacrylate
PVC	polyvinyl chloride (polymer)
QCL	quantum cascade laser
RF	random forest
RGB	red, green, blue
SNR	signal-to-noise ratio
SVM	support vector machine
TPR	true positive rate
VIS	visual (visual spectral range)
YR	yellow rust

Author details

Ran Aharoni[1*], Asaf Zuck[2], David Peri[3] and Shai Kendler[1,4]

1 Department of Environmental Physics, Israel Institute for Biological Research, Ness-Ziona, Israel

2 Department of Chemical Physics, Israel Institute for Biological Research, Ness-Ziona, Israel

3 Independent Scientist, Israel

4 Faculty of Civil and Environmental Engineering, Water and Agricultural Engineering, Technion—Israeli Institute of Technology, Haifa, Israel

*Address all correspondence to: tenzoran@gmail.com

References

[1] Van Neste CW, Senesac LR, Thundat T. Standoff spectroscopy of surface adsorbed chemicals. Analytical Chemistry. 2009;**81**(5):1952-1956

[2] Aharoni R, Klymiuk V, Sarusi B, Young S, Fahima T, Fishbain B, et al. Spectral light-reflection data dimensionality reduction for timely detection of yellow rust. Precision Agriculture. 2020;**22**:267-286. DOI: 10.1007/s11119-020-09742-2

[3] Staymates JL, Staymates ME, Lawrence J. The effect of reusing wipes for particle collection. International Journal for Ion Mobility Spectrometry. 2015;**19**(1):41-49

[4] Garcia-Reyes JF, Harper JD, Salazar GA, Charipar NA, Ouyang Z, Cooks RG. Detection of explosives and related compounds by low-temperature plasma ambient ionization mass spectrometry. Analytical Chemistry. 2010;**83**(3): 1084-1092

[5] Justes DR, Talaty N, Cotte-Rodriguez I, Cooks RG. Detection of explosives on skin using ambient ionization mass spectrometry. Chemical Communications. 2007;**21**:2142-2144

[6] He J, Tang F, Luo Z, Chen Y, Xu J, Zhang R, et al. Air flow assisted ionization for remote sampling of ambient mass spectrometry and its application. Rapid Communications in Mass Spectrometry. 2011;**25**(7):843-850

[7] Serrano J, Moros J, Sánchez C, Macías J, Laserna JJ. Advanced recognition of explosives in traces on polymer surfaces using LIBS and supervised learning classifiers. Analytica Chimica Acta. 2014;**806**:107-116

[8] Tourne M. Developments in explosives characterization and detection. Journal of Forensic Research. 2014;**S12**:002. DOI: 10.4172/2157-7145. S12-002

[9] Li H, Tripp CP. Detection of Bacillus globigii spores using a fourier transform infrared-attenuated total reflection method. Applied Spectroscopy. 2008; **62**(9):963-967

[10] Andrew Chan KL, Kazarian SG. Detection of trace materials with Fourier transform infrared spectroscopy using a multi-channel detector. The Analyst. 2006;**131**(1):126-131

[11] Harig R, Braun R, Dyer C, Howle C, Truscott B. Short-range remote detection of liquid surface contamination by active imaging Fourier transform spectrometry. Optics Express. 2008;**16**(8):5708

[12] Braun R, Harig R. Stand-off identification and mapping of liquid surface contaminations by passive hyperspectral imaging. Chemical, Biological, Radiological, Nuclear, and Explosives Sensing XIV. 2013;**8710**: 871004

[13] Dobroiu A, Sasaki Y, Shibuya T, Otani C, Kawase K. THz-wave spectroscopy applied to the detection of illicit drugs in mail. Proceedings of the IEEE. 2007;**95**(8):1566-1575

[14] Wang S, Ferguson B, Mannella C, Abbott D, Zhang X-C. Powder detection using THz imaging. In: Summaries of Papers Presented at the Quantum Electronics and Laser Science Conference. America: Opt. Soc.; 2002. p. 44

[15] Patel CKN. From CO_2 lasers to quantum cascade lasers—A saga of high power infrared lasers. Journal of Laser Applications. 2010;**21**(4):224-238

[16] Tittel FK, Richter D, Fried A. Mid-infrared laser applications in spectroscopy. Solid-State Mid-Infrared Laser Sources. 2007;**516**(7):458-529

[17] Chen X, Guo D, Choa F-S, Wang C-C, Trivedi S, Snyder AP, et al. Standoff photoacoustic detection of explosives using quantum cascade laser and an ultrasensitive microphone. Applied Optics. 2013;**52**(12):2626

[18] Patel CKN, Barron-Jimenez R, Dunayevskiy I, Tsvid G, Lyakh A. Two wavelength operation of an acousto-optically tuned quantum cascade laser and direct measurements of quantum cascade laser level lifetimes. Applied Physics Letters. 2017;**110**:031104. DOI: 10.1063/1.4974214

[19] Craig IM, Taubman MS, Lea AS, Phillips MC, Josberger EE, Raschke MB. Infrared near-field spectroscopy of trace explosives using an external cavity quantum cascade laser. Optics Express. 2013;**21**(25):30401

[20] Phillips MC, Bernacki BE. Hyperspectral microscopy of explosives particles using an external cavity quantum cascade laser. Optical Engineering. 2012;**52**(6):061302

[21] Goyal AK, Wood D, Lee V, Rollag J, Schwarz P, Zhu L, et al. Laser-based long-wave-infrared hyperspectral imaging system for the standoff detection of trace surface chemicals. Optical Engineering. 2020;**59**(09):1

[22] Pavlenko AA, Maksimenko EV, Chernyshova LV. Stand-off detection of HMX traces by active spectral imaging with a tunable CO_2 laser. Quantum Electronics. 2014;**44**(4):383-386

[23] Kendler S, Aharoni R, Cohen S, Raich R, Weiss S, Levy H, et al. Non-contact and non-destructive detection and identification of *Bacillus anthracis* inside paper envelopes. Forensic Science International. 2019;**301**:e55-e58

[24] Yang J, Messinger DW. Bloodstain detection and discrimination impacted by spectral shift when using an interference filter-based visible and near-infrared multispectral crime scene imaging system. Optical Engineering. 2018;**57**(03):1

[25] Stenzel O. The Physics of Thin Film Optical Spectra: An Introduction (Springer Series in Surface Sciences). 2nd ed. Vol. 44. Cham: Springer International Publishing; 2016. 1-352 p

[26] Yariv A. Quantum Electronics. 3rd ed. New York: Wiley; 2020. Available from: Wiley.com

[27] Kovalenko SA. Descartes-Snell law of refraction with absorption. Semiconductor Physics, Quantum Electronics and Optoelectronics. 2001; **4**(3):214-218

[28] Mugele RA, Evans HD. Droplet size distribution in sprays. Industrial and Engineering Chemistry. 1951;**43**(6): 1317-1324

[29] Kippax P, Parkin S, Tuck C. Particle size characterisation of agricultural sprays using laser diffraction. ILASS-Europe. 2002;**6**(2):2-7

[30] Patterson MA, Reitz RD. Modeling the effects of fuel spray characteristics on diesel engine combustion and emission. In: SAE Technical Papers. SAE International; 1998

[31] Mishchenko MI. Maxwell's equations, radiative transfer, and coherent backscattering: A general perspective. Journal of Quantitative Spectroscopy and Radiation Transfer. 2006;**101**(3):540-555

[32] Wriedt T. Mie theory: A review. In: Springer Series in Optical Sciences. Berlin, Heidelberg: Springer; 2012. pp. 53-71

[33] Young AT. Rayleigh scattering. Physics Today. 1982;**35**(1):42-48

[34] van de Hulst HC. Light Scattering by Small Particles (Dover Books on

Physics). 2nd ed. New York: Dover Publications; 1981. 496 p

[35] Bohren CF. Scattering by a sphere and reflection by a slab: Some notable similarities. Applied Optics. 1988;**27**(2): 205

[36] Aharoni R, Elisha S, Peri D, Kendler S. Liquid trace detection and identification by spectral reflectance model. Optical Engineering. 2019; **58**(07):1

[37] Herschel W. XIV. Experiments on the refrangibility of the invisible rays of the sun. Philosophical Transactions. Royal Society of London. 1800;**90**: 284-292

[38] Herschel W. XIII. Investigation of the powers of the prismatic colours to heat and illuminate objects; with remarks, that prove the different refrangibility of radiant heat. To which is added, an inquiry into the method of viewing the sun advantageously, with telesco. Philosophical Transactions. Royal Society of London. 1800;**90**: 255-283

[39] Herschel W. XV. Experiments on the solar, and on the terrestrial rays that occasion heat; with a comparative view of the laws to which light and heat, or rather the rays which occasion them, are subject, in order to determine whether they are the same, or different. Philosophical Transactions. Royal Society of London. 1800;**90**:293-326

[40] Adams MJ. Chemometrics in Analytical Spectroscopy (RSC Analytical Spectroscopy Series). Cambridge: Royal Society of Chemistry; 2007

[41] Manley M. Near-infrared spectroscopy and hyperspectral imaging: Non-destructive analysis of biological materials. Chemical Society Reviews. 2014;**43**(24): 8200-8214

[42] Sendin K, Williams PJ, Manley M. Near infrared hyperspectral imaging in quality and safety evaluation of cereals. Critical Reviews in Food Science and Nutrition. 2018;**58**(4):575-590

[43] Qin J, Lu R. Measurement of the optical properties of fruits and vegetables using spatially resolved hyperspectral diffuse reflectance imaging technique. Postharvest Biology and Technology. 2008;**49**(3):355-365

[44] Lammertyn J, Peirs A, De Baerdemaeker J, Nicolaï B. Light penetration properties of NIR radiation in fruit with respect to non-destructive quality assessment. Postharvest Biology and Technology. 2000;**18**(2):121-132

[45] Norouzi S, Sadeghi M, Liaghat A, Tuller M, Jones SB, Ebrahimian H. Information depth of NIR/SWIR soil reflectance spectroscopy. Remote Sensing of Environment. 2021;**256**: 112315

[46] Park B, Yoon S-C, Windham WR, Lawrence KC, Kim MS, Chao K. Line-scan hyperspectral imaging for real-time in-line poultry fecal detection. Sensing and Instrumentation for Food Quality and Safety. 2011;**5**(1):25-32

[47] Goetz AFH, Vane G, Solomon JE, Rock BN. Imaging spectrometry for Earth remote sensing. Science (80-). 1985;**228**(4704):1147-1153

[48] Ben-Dor E, Patkin K, Banin A, Karnieli A. Mapping of several soil properties using DAIS-7915 hyperspectral scanner data—A case study over soils in Israel. International Journal of Remote Sensing. 2002;**23**(6): 1043-1062

[49] Brook A, Dor E. Ben. Supervised vicarious calibration (SVC) of hyperspectral remote-sensing data. Remote Sensing of Environment. 2011; **115**(6):1543-1555

[50] Manolakis D, Shaw G. Detection algorithms for hyperspectral imaging applications. IEEE Signal Processing Magazine. 2002;**19**(1):29-43

[51] Bioucas-Dias JM, Plaza A, Dobigeon N, Parente M, Du Q, Gader P, et al. Hyperspectral unmixing overview: Geometrical, statistical, and sparse regression-based approaches. IEEE Journal of Selected Topics in Applied Earth Observations and Remote Sensing. 2012;**5**(2):354-379

[52] Kendler S, Ron I, Cohen S, Raich R, Mano Z, Fishbain B. Detection and identification of sub-millimeter films of organic compounds on environmental surfaces using short-wave infrared hyperspectral imaging: Algorithm development using a synthetic set of targets. IEEE Sensors Journal. 2019; **19**(7):2657-2664

[53] Hollis J, Raich R, Kim J, Fishbain B, Kendler S. Foreground signature extraction for an intimate mixing model in hyperspectral image classification. In: ICASSP 2020—2020 IEEE International Conference on Acoustics, Speech and Signal Processing (ICASSP). IEEE; 2020. pp. 4732-4736

[54] Sengupta A, Bandyopadhyay A, Barat RB, Gary DE, Federici JF. THz reflection spectroscopy of C-4 explosive and its detection through interferometric imaging. In: Hwu RJ, Linden KJ, editors. Proc. SPIE. Terahertz and Gigahertz Electronics and Photonics V. Vol. 6120. 2006. p. 61200A. DOI: 10.1117/12.641727

[55] Deutsch ER, Kotidis P, Zhu N, Goyal AK, Ye J, Mazurenko A, et al. Active and passive infrared spectroscopy for the detection of environmental threats. Advanced Environmental, Chemical, and Biological Sensing Technologies XI. 2014;**9106**:91060A

[56] Kendler S, Aharoni R, Cohen S, Raich R, Weiss S, Levy H, et al. Non-contact and non-destructive detection and identification of *Bacillus anthracis* inside paper envelops. Forensic Science International. 2019;**301**:e55-e58. DOI: 10.1016/j.forsciint.2019.05.007

[57] Nelson MP, Shi L, Zbur L, Priore RJ, Treado PJ. Real-time short-wave infrared hyperspectral conformal imaging sensor for the detection of threat materials. In: Fountain AW, editor. Chemical, Biological, Radiological, Nuclear, and Explosives (CBRNE) Sensing XVII. International Society for Optics and Photonics; 2016. p. 982416

[58] Zhi T, Pires BR, Narasimhan SG. Multispectral imaging for fine-grained recognition of powders on complex backgrounds. In: 2019 IEEE/CVF Conference on Computer Vision and Pattern Recognition (CVPR). 2019. pp. 8691-8700. DOI: 10.1109/CVPR.2019.00890

[59] Kampouris C, Zafeiriou S, Ghosh A, Malassiotis S. Fine-grained material classification using micro-geometry and reflectance. In: Leibe B, Matas J, Sebe N, Welling M, editors. Computer Vision—ECCV 2016. Lecture Notes in Computer Science. Vol. 9909. Cham: Springer; 2016. pp. 778-792. DOI: 10.1007/978-3-319-46454-1_47

[60] Breiman L, Friedman J, Stone C, Olshen R. Classification and Regression Trees. Routledge; 1984. DOI: 10.1201/9781315139470

[61] Breiman L. Bagging predictors. Machine Learning. 1996;**24**(2):123-140

[62] Breiman L. Random forests. Machine Learning. 2001;**45**(1):5-32. DOI: 10.1023/A:1010933404324

[63] Reiss A, Jørgensen LN. Biological control of yellow rust of wheat (*Puccinia striiformis*) with Serenade® ASO (*Bacillus subtilis* strain QST713). Crop Protection. 2017;**93**:1-8

[64] Basandrai A, Sharma B, Basandrai D. Efficacy of triazole fungicides for the integrated management of yellow rust, leaf rust and powdery mildew of wheat. Plant Disease Research. 2013;**28**(2): 135-139

[65] Klymiuk V, Yaniv E, Huang L, Raats D, Fatiukha A, Chen S, et al. Cloning of the wheat Yr15 resistance gene sheds light on the plant tandem kinase-pseudokinase family. Nature Communications. 2018;**9**:3735. DOI: 10.1038/s41467-018-06138-9

[66] A Global Wheat Rust Monitoring System. Available from: RustTracker.org [cited 2021 Sep 13]

Section 2

Spectrophotometry for Evaluation of Colour

Chapter 4

Spectroscopy and Spectrophotometry: Principles and Applications for Colorimetric and Related Other Analysis

Murali Dadi and Mohd Yasir

Abstract

Spectrophotometry and different types of spectroscopy are the technique that involved in identifying and quantifying the amount of a known substance in an unknown medium. Spectroscopy is the most convenient method for analysis of unknown samples both qualitatively and quantitatively with a good percentage of accuracy. Different types of spectroscopic and spectrophotometric techniques are very helpful in analyzing the samples even at sub-ppm level particularly in the field of scientific research. These techniques based on the simple principle that the amount of specific radiation i.e. ray or light (photon) absorbed or reflected by the sample relative to the intensity of the incident ray/light at a particular wavelength. These techniques are using, for analyzing purity, % content in mixture, type of reactions/chemical interactions occur/absorption or reflectance of color for a colored substances/solutions are detectable and quantitatively determinable quantitative determination. Most of the scientists have been using different spectroscopic and spectrophotometric techniques like Infrared spectroscopy, Raman spectroscopy, X-ray fluorescence and UV VIS spectrophotometry etc., which are playing an important role in the identification and characterization of substances, apart from this the atomic absorption spectroscopy and atomic emission spectroscopy are also being used for quantitative measurement of different substances or elements.

Keywords: spectroscopy, spectrophotometry, UV-visible spectroscopy, infrared spectroscopy, Raman spectroscopy, X-ray fluorescence

1. Introduction

Every compound, that is present in the nature, has a property to absorb, transmit, or reflect light (electromagnetic radiation) at a certain wavelength. This property of the compounds, helps to measure quantitatively by using spectrophotometric techniques. Spectrophotometry is a technique which deals with the measurement of the interaction of light with materials. When light falls on a material that can be reflected, transmitted, scattered, or absorbed, and at the same time the material on which light has fallen can emit absorbed light with different frequency. This is due to the gained energy from the light (e.g., electroluminescence) or due to its temperature (incandescence) [1]. Different types of spectroscopy and spectrophotometry is well known and widely used technique to identify and quantify

IntechOpen

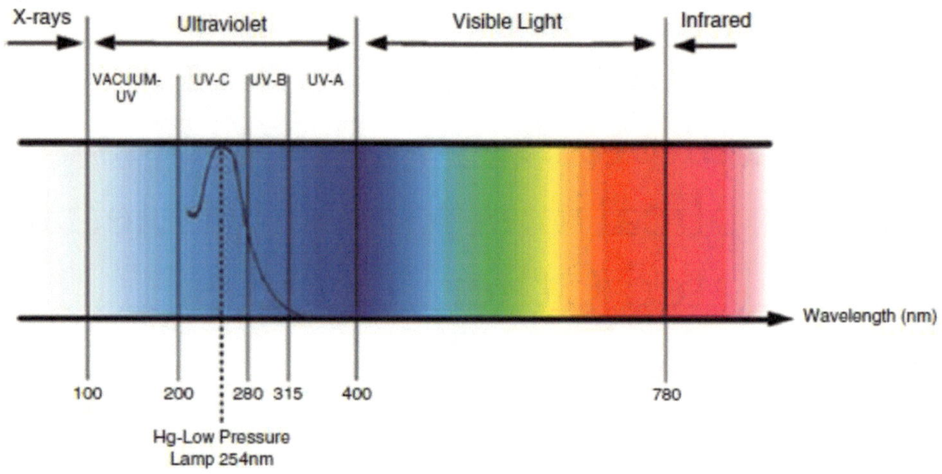

Figure 1.
Electromagnetic Spectrum.

compounds in the field of research as well as in the industrial and chemical laboratories. For example, in chemistry, and pharmacy, UV-visible spectrophotometry is a basic technique to analyze the samples based on the application of the Beer-Lambert-Bouguer Law. In biochemistry and molecular biology, spectrophotometric analysis is essential for determining biomolecule concentration of a solution and is employed ubiquitously for determining the concentration of DNA, RNA, or protein [2]. In clinical laboratories both manual and automated spectrophotometric are extensively used for the determination of blood, urine, and body fluid samples [3].

Several types of spectroscopic and spectrophotometric methods are applied to analyze the samples. Among them, there are two primary methods which are highly utilized; absorption spectrophotometry, which is based on the absorption of radiation at specific wavelength of light to get absorption spectrum, and UV-visible spectrophotometry, which is apprehensive with the reflectance of specific spectra of a given material within UV and visible range of electromagnetic radiation spectrum (**Figure 1**) [4].

2. Principle, instrumentation and applications of various spectrophotometric techniques

A spectroscopic/spectrophotometric instrument basically consists of four important components: a light/radiation source, a collimator, a monochromator, and a detector. The monochromator comprehends a fixed entrance slit, a dispersing element such as a prism or a diffraction grating, and a moving exit slit (**Figure 2**) [6].

2.1 UV-visible spectrophotometry

2.1.1 Principle

Law of absorption is the basic principle of UV-visible spectrophotometry. This law discusses the relation between thickness of the absorbing material and the concentration of the sample solution, which is popularly known as Beer-Lambert law or simply Beer's law. This law states that the amount of light absorbed is proportional to the concentration of the absorbing substance and to the thickness of the absorbing material [7].

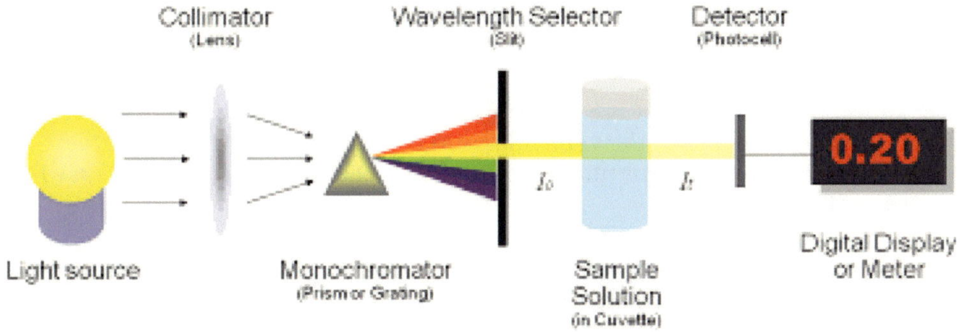

Figure 2.
Basic instrumentation of spectrophotometer [5].

$$\log_{10} = \frac{I_0}{I} = abC \tag{1}$$

where I_0 = the intensity of the incident light, I = the intensity of the transmitted light, a = absorption, b = the absorbing thickness, C = the concentration of the absorbing material.

2.1.2 Instrumentation

The UV-visible spectrophotometer consists of a light source, sample holders, a monochromator, and a detector [8].

Light source: Hydrogen lamp and the deuterium lamps are used as UV light source, whereas for visible source tungsten filament lamp is the most used.

Sample holders: In UV and Visible ranges, cuvettes are used as sample holders, which are made from quartz or ordinary glass. Generally, in the UV region quartz or silica cell are used, whereas in the visible region glass cell are used and these cuvettes have a standard path length is usually 1 cm.

Monochromators: A monochromator converts polychromatic radiation into monochromatic radiations by which the wavelengths of these radiations translate into very narrow bands.

Detectors: Photovoltaic cells, phototubes and photomultiplier are commonly used detectors in the UV and visible range [7]. The following block diagram (**Figure 3**) shows main parts of UV-Visible spectrophotometer [7].

2.1.3 Applications of UV-visible spectrophotometry

a. Pharmaceutical analysis: UV-visible Spectrophotometry has been widely used technique in the determination of drug concentration in pharmaceutical analysis.

   For example, this technique is used in the determination of etravirine in bulk and pharmaceutical formulations. The spectrum of etravirineis shown below. This is acting as an anti-viral drug, and it showed the maximum absorption at 414 nm

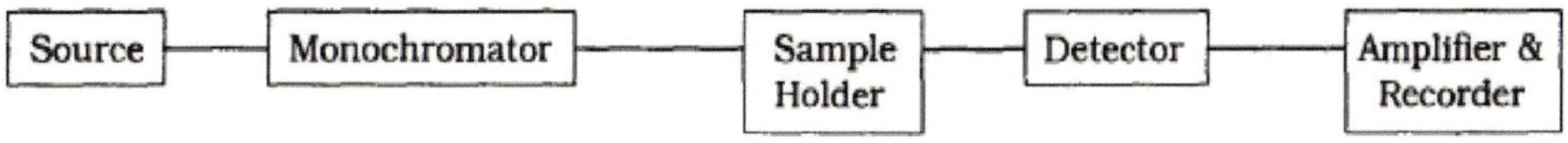

Figure 3.
Block diagram of UV-visible spectrophotometer.

(visible range) by reacting with NaOH and 1,2-napthaquinone 4-sulphonate. The details are shown in the following (**Figure 4**) [9].

b. Vaporization studies of low-volatile compounds: UV-visible spectrophotometer is involved in the determination of vaporization of low-volatile compounds in the vapor phase located above the sample in the condensed state [10].

c. UV-visible spectroscopy is also used in the identification of pure analytes which are not subjected to decomposition, particularly this is used for identification of nucleic acids. This study has an advantage to identify the newly found genetic materials in various microbes and other species. The following (**Figure 5**) shows absorption of DNA in UV light [11].

d. Quantification and identification of organic compounds has been achieved by using UV-visible spectroscopy technique. This technique is very helpful for the analyzing of newly developed drugs in the pharmaceutical industries [12].

e. UV-visible spectrophotometry is widely used technique in biochemistry for the determination of micromolar concentrations of substances in blood, urine, and other body fluids. It is also used both for the determination of species and for studying biochemical processes [13].

f. UV-vis spectroscopy also involved in the assessing the Color Index of Transformer Insulating Oil (**Figure 6**) [14].

g. A case study has been conducting on combined UV-visible spectroscopy and chemometrics to determine the interaction of human serum albumin (HAS) and gold nanoparticles (AuNPs). The data which has been recovered from the UV-visible spectroscopy and chemometrics about protein (HAS) interaction with nanoparticles (AuNPs) were apply to the thermodynamic, kinetic and structural parameters to establish the evolution of protein nanoconjugate (**Figure 7**) [15].

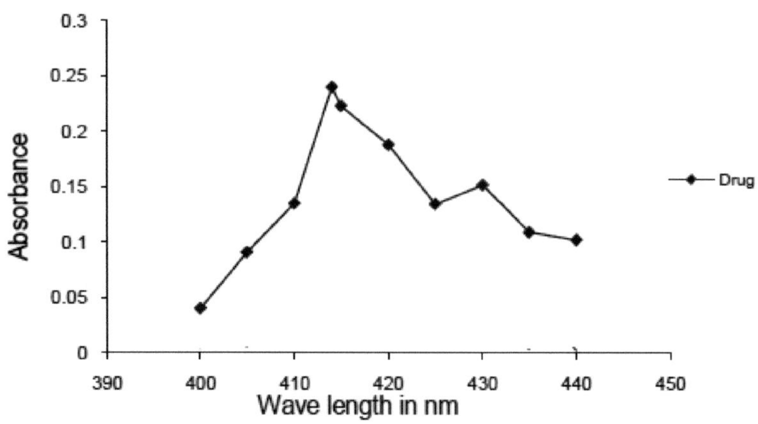

Figure 4.
UV-visible spectrum of etravirine.

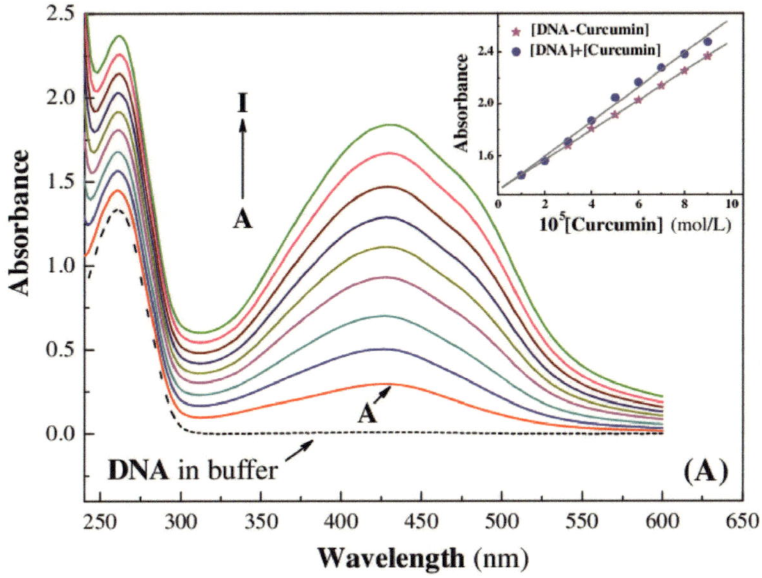

Figure 5.
UV absorption spectrum of DNA.

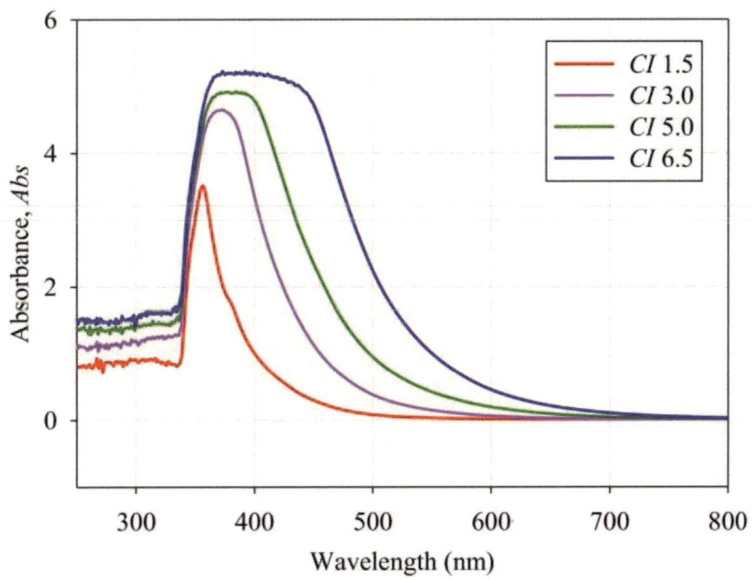

Figure 6.
Optical absorbance spectrums of four oil samples with different CI based on ASTM D 1500 after applying ND filter [14].

2.2 Infrared spectrophotometry

2.2.1 Principle

When a molecule absorbs the light of higher wavelength rather than UV and visible, then there is a possibility of vibrational transitions in the molecules. These vibrational transitions of the molecules lead to the formation of an IR spectrum.

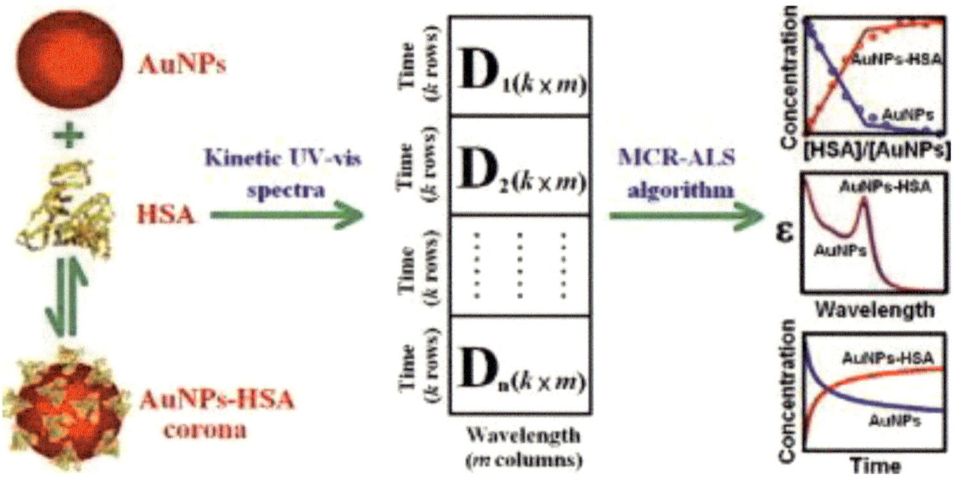

Figure 7.
Interaction human serum albumin (HAS) with citrate-capped gold nanoparticles (AuNPs) [15].

These vibrational transitions due to occurrence of electronic transitions when a given substance absorbs light energy.

2.2.2 Instrumentation

Like UV-visible spectrophotometer, IR spectrometer also consists of a light source, a sample holder, a monochromator, and a detector [7].

Light source: Xenon and tungsten lamps are typically used as light sources in the near IR-region [16].

Sample holders: In IR region quartz cuvettes usually used as sample holders.

Monochromator: Gratings are used as monochromators in the IR region.

Detectors: Commonly used detectors in the IR spectrophotometry are indium gallium arsenide (InGaAs) semiconductor materials.

2.2.3 Methodology

The radiation energy is the main element in the IR spectrophotometer. It acts as a function of wavelength and provides energy to reach a maximum at a wavelength (μm) equal to 2897/T, where T is the absolute temperature (K). This radiation energy is usually providing a short-wavelength limit of the spectrum (~2 μm) and it decreases as the wavelength gets longer [17]. The radiation energy coming from the source falls on to the samples, causing the excitations followed by molecular vibrations in the sample by which it can give IR spectrum that can be detected by detectors. The detectors can change the radiation energy into an electrical signal that can be amplified and processed to yield a spectrum. The spectrum gives information about various functional present in the sample.

2.2.4 Applications of IR spectroscopy

a. Identification of compounds: IR spectroscopy assists in finding out various chemical compounds and functional groups in organic molecules, such as aliphatic, aromatic, saturated and unsaturated hydrocarbons, amino acids, ether and hydroxyl groups, halogens, nitrogen, phosphorous, silicon, sulfur-oxy compounds etc.

The aliphatic and aromatic hydrocarbons can be analyzed by C–H and C–C stretching and bending vibrations, most of these vibrations are unique for each molecule, and are generally described as skeletal vibrations. The C═C–C bond in the ring structure of aromatic compounds is diagnosed by characteristic stretching and bending vibrations [18].
For example, the spectrum of 1-hexene shows characteristic absorptions of a double bond. The C–H stretch at 3080 cm^{-1} corresponds to the alkene ═C–H bonds. The absorption at 1642 cm^{-1} results from stretching of the C═C double bond. The diagram represents the IR Spectra of 1-hexene (**Figure 8**) [18].

b. IR spectroscopy successfully involved in the characterization of nano particles particularly in the study of physicochemical characteristics of drug nanocarriers and in the identification of

c. Functional groups on the surface of the developed nano particles which are involved in drug targeting system [19].

d. IR spectroscopy takes an important role in the surface biology research to study the surface interaction of drugs, antibodies with cell surface proteins and other biological molecules, which helps in understanding how to optimize sensitivity in between the interacted molecules [20].

e. Rate of reactions: infrared spectra give wonderful indication for many functional groups. Thus, enzymatic reactions involving these functional groups—either these groups are consumed or generated in the enzymatic reaction—can be assayed with the help of infrared spectroscopy. For example, the enzymatic activity of the pyruvate kinase has been studied with its substrate phosphoenol pyruvate, which gave a characteristic spectrum for understanding how the substrate consumed and the product formed. The details are shown in the following (**Figure 9**) [21].

f. Interaction between molecules: polypeptide chains form inter chain hydrogen bonds. So do the two stands of DNA. Hydrogen bonds have been studied very profitably using infrared spectroscopy.

g. Infrared spectroscopy is an important tool in structural determination of minerals. For example, in one of the case studies which has been done on the alternations

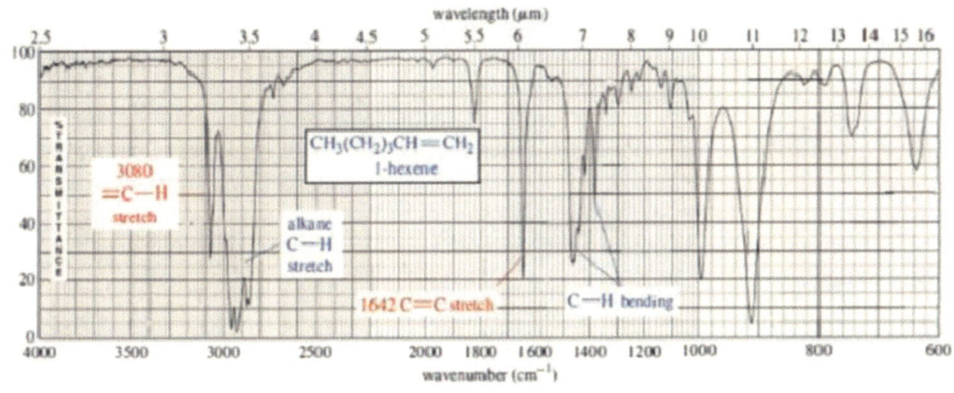

Figure 8.
IR spectra of 1-hexene.

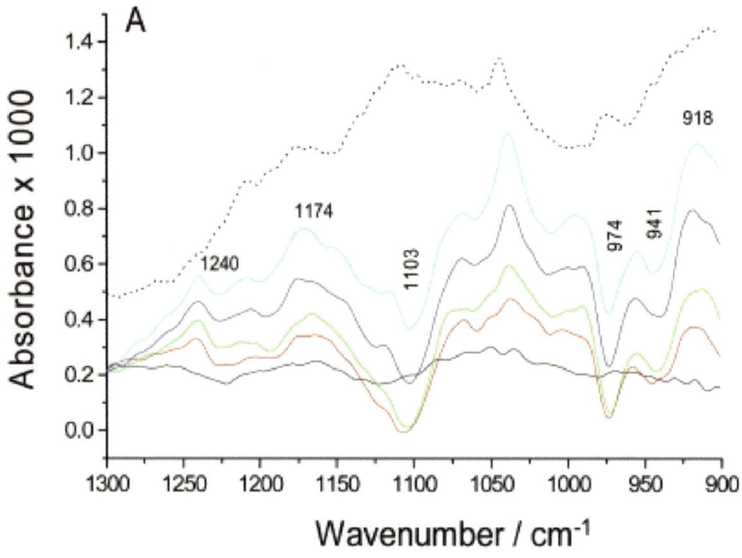

Figure 9.
Enzymatic reaction of PK. (A) Series of overlaid spectra (solid lines) of infrared absorbance changes upon PEP and ADP addition to PK, observed for 30 min.

of chondrule in NWA 2086 CV3 meteorite by using IR spectroscopy along with optical microscopy and electron microprobe. This study revealed that the alternations have brought changes in intensity and wavelength positions of olivine peaks with the advancement of alteration and related Fe/Mg substitution inward of the chondrule and also it was identified that there is a good correlations between Fo% composition and positions of 830 and 860 cm^{-1} IR peaks [22].

2.3 Fourier transform infrared spectroscopy (FTIR)

2.3.1 Principle

FTIR works on the principle of IR spectroscopy. Nevertheless, the instrumentation is different from IR spectroscopy.

2.3.2 Instrumentation

FTIR spectrometer consists of a light source, a sample holder, a monochromator, and a detector which are like that of IR spectrophotometer, but the major difference is an interferometer, which makes this instrument highly advanced than normal IR spectrophotometer. The interferometer specially consists of a compensator plate, a beam splitter, a fixed mirror, and a scanning mirror, which are connected to a detector. The advantages of FTIR over the existing dispersive infrared instrument are spectral quality, data collection speed, reproducibility of data, and ease of maintenance and use. Instrumentation of the FTIR is shown in the following (**Figure 10**) [23].

2.3.3 Applications

Fourier transform infrared (FTIR) spectroscopy is a powerful analytical tool in identifying chemical constituents and elucidating structures in various forms in real-world samples.

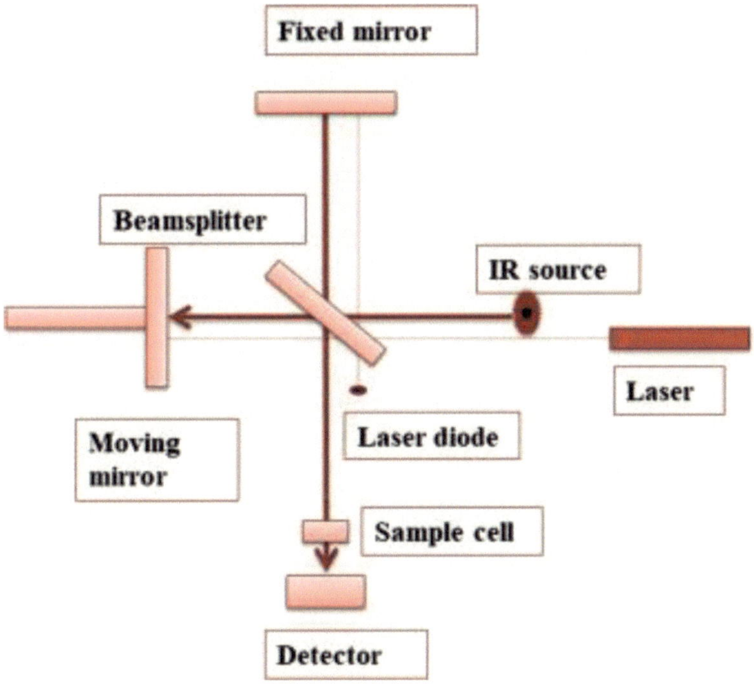

Figure 10.
Schematic diagram of FTIR spectrometer.

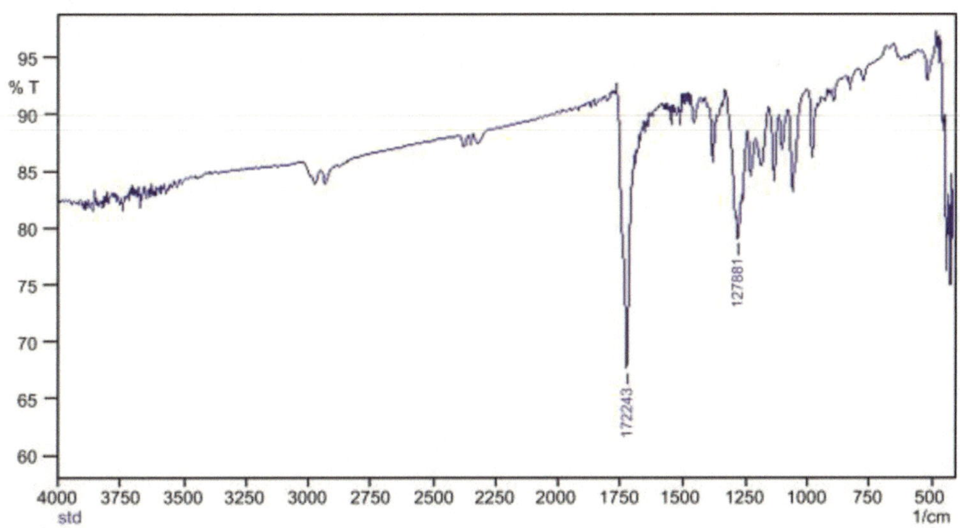

Figure 11.
FTIR spectrum of poly-3-hydroxybutyrate (PHB).

a. FTIR has been used for the characterizing the unpredictability in fuel stability of various biodiesel and antioxidant samples. This can be achieved by identifying the presence of various organic and inorganic compounds in the sample through the FTIR spectrum [24].

b. FTIR has been used for the identification of functional groups in polymers and co-polymers. For examples, FTIR used in the identification of functional groups in one of the co-polymers i.e., poly-3-hydroxybutyrate (PHB), which shows peaks at 1724 cm^{-1} and 1279 cm^{-1} corresponding to C–O stretching and the adsorption band respectively in the ester group. The following (**Figure 11**) shows the C–O stretching and the adsorption band of poly-3-hydroxybutyrate (PHB) [25].

c. Fourier transform infrared (FTIR) spectroscopy with attenuated total reflectance (ATR) accessory was used in forensic analysis to get biochemical information on postmortem interval estimation based on pericardial fluids of rabbit.

d. Fourier transform infrared (FTIR) spectroscopy with attenuated total reflectance (ATR) has also been used for the assessment of the immobilization of active substances in the matrix of biomedical materials. The presence of Sparfloxacin on the surface of the biomaterial was studied by using FTIR method. In this study the Sparfloxacin has shown many absorption bands in FTIR spectrum, which reveals the information that the successful binding of an antibiotic with bacterium (**Figure 12**) [26].

2.4 Raman spectroscopy

2.4.1 Principle

Raman spectroscopy based on the scattering of light, which was described by C.V. Raman in 1928 through his outstanding study i.e., Raman effect. According to Raman effect when a certain frequency of monochromatic radiations incident on a sample, the incident light is scattered through interaction with vibrating sample molecules, the frequency of scattered light is different from that of the incident light. It is based on the inelastic scattering of incident radiation [27]. In Raman spectroscopy, when a monochromatic radiation strikes the sample, it scatters in all directions after its interaction with sample molecules. Much of this scattered radiation elastically that constitutes Rayleigh scattering. Only a small fraction of scattered radiation in-elastically scattered that constitutes Raman

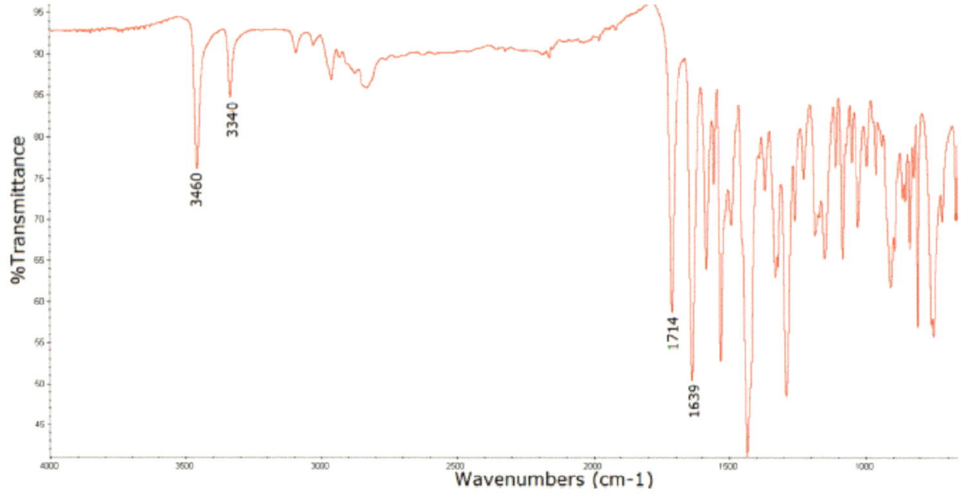

Figure 12.
FTIR spectrum of Sparfloxacin.

scattering. Usually, in Raman spectrum there is an appearance of Stokes lines, this is due to the frequency of incident radiation is higher than frequency of scattered radiation. On the other hand, there is an appearance of anti-Stokes lines in Raman spectrum, when the frequency of incident radiation is lower than frequency of scattered radiation occurs.

2.4.2 Instrumentation

Instrumentation for modern Raman spectroscopy consists of following components:

a. Light source

b. Prism or grating

c. Detectors

a. Light source: during 1960s Mercury lamps were used as light source. From late 1960s various kinds of lasers were started to use as light sources as they provide stable and intensive beam of light in Raman spectrophotometers that makes the Raman spectroscopy more versatile than ever. Long wavelength sources such as diode or Nd:YAG lasers are preferred as they have advantage and can be operated at much higher power without causing photodecomposition of sample and eliminates or reduces fluorescence in most cases [28].

b. Prism or grating: prism or grating uses in dispersive Raman spectrophotometer while Michelson interferometer uses in non-dispersive Raman spectrophotometer. The grating monochromators are used to separate relatively weak Raman lines from intense Rayleigh scattered radiations.

c. Detectors: in earlier models of dispersive Raman spectrophotometers, thermoelectrically cooled photomultiplier tubes and photodiode array detectors were used. Now a days, due to the advancement in instrumentation technology, more sensitive charge transfer devices (CTDs) such as charge-coupled devices (CCDs) and charge-injection devices (CIDs) are in usage. These devices act as a detector and use in the form of arrays .

The (**Table 1**) one shows some common laser sources which are generally use in Raman Spectroscopy [29] and (**Figure 13**) shows the instrumentation of Raman Spectrophotometer [30].

Laser type	Wavelength, nm
Argon ion	488.0 or 514.5
Krypton ion	530.9 or 647.1
Helium-neon	632.8
Diode	785 or 830
Nd-YAG	1064

Table 1.
Some common laser sources for Raman spectroscopy.

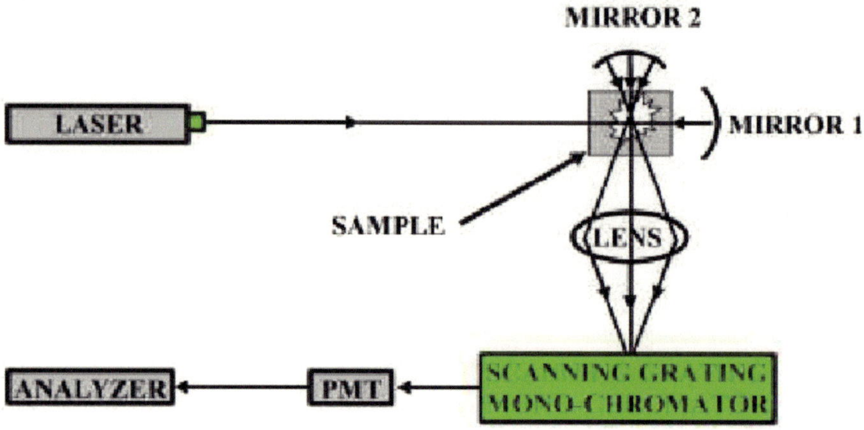

Figure 13.
Instrumentation of Raman spectrometer.

2.4.3 Raman spectrum

Typically Raman spectrum is plotted signal intensity vs. Raman shift (Raman shift, in cm^{-1} = energy of photon in – energy of photon out). **Figure 14** shows the Raman spectrum.

2.4.4 Applications

a. Cell therapy: Raman spectroscopy involved in the development of cell therapies. Cell therapy is typically done by introducing a living cell into the patients to cure many degenerative and deadly diseases. Raman spectroscopy plays an important role in basic understanding of biochemical and functional characteristics of therapeutic cells, in manufacturing them and to effectively implement them for therapy [31].

b. Molecular diagnosis of cervical cancer: Raman spectroscopy is powerful tool in biochemical investigation, particularly in identifying and characterizing the

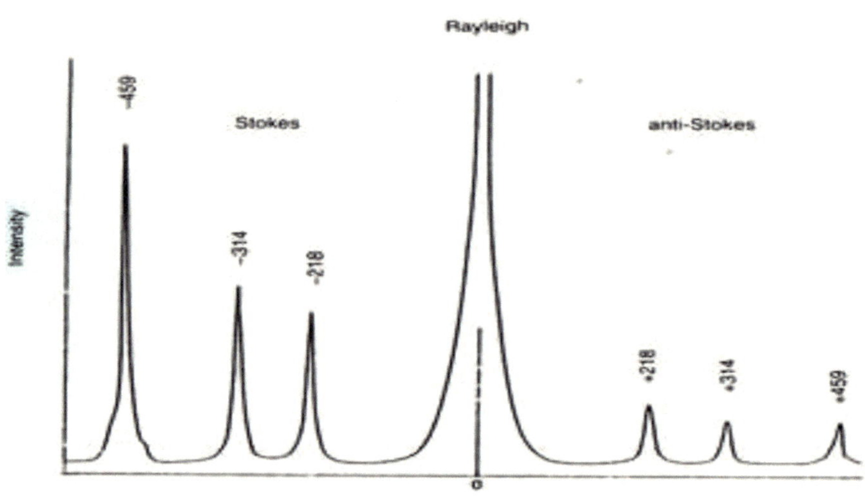

Figure 14.
Raman Spectrum.

structure of biomolecules, cells, and tissues. This approach is very helpful for the scientists to diagnose the malignant neoplasm which leads to the cervical cancer [32].

c. In agriculture, food and biosystems: Raman spectroscopy is often used for early detection of plant diseases. For example, for the early detection of Citrus Huanglongbing and rose rosette disease (RRD). Raman spectroscopy combined with chemometric analysis provides the information about the effect of Tomato yellow leaf curl Sardinia virus (TYLCSV) and Tomato spotted wilt virus (TSWV) on tomato. Raman spectroscopy is one of the effective tools in identifying the food adulteration. For examples Raman spectroscopy combined with chemometric methods are used in the determination of butter adulteration. SERS is used in the determination of toxic effect of mycotoxin deoxynivalenol in corn, kidney beans, and oats [33].

d. Raman spectroscopy like IR spectroscopy used in chemistry to identify the chemical structures by providing the fingerprint region in the spectrum and it is involved in the functional group identification.

e. Raman spectroscopy is one of the powerful techniques in biopharmaceutical industry. It helps in measurement and analysis of particle size while preparation of the drug and it involved in the identification of contaminants which are coming out from the pipes, valves, bags, filters, etc. during formulation the drug in pharmaceutical industry.

f. Raman spectroscopy involved in the identification protein structure, protein glycosylation, protein stability and aggregation, and also in protein formulations and identity testing which are main important aspects in the biopharmaceutical manufacturing [34].

2.5 Spectro-fluorometry

2.5.1 Principle

The phenomenon, where a molecule after absorbing radiations emits radiation of a longer wavelength, is known as fluorescence. When a compound absorbs radiation, it goes to the exited level and thereafter comes to the ground level either in one step by emitting radiation of the same wavelength that it absorbed or in a stepwise manner by emitting quanta of radiation corresponding to each energy step with longer wavelength. This phenomenon leads to the formation of fluorescence spectra. Fluorescence is an extremely short-lived phenomenon (10^{-7} s or less) and therefore can provide information about events which takes less than 10^{-7} s. Fluorometry also follow the Beer-Lambert law in its working principle.

2.5.2 Instrumentation

Fluorometry is an important analytical tool for the determination of extremely small concentrations of substances which exhibit fluorescence.

2.5.3 Instrumentation

The instrumentation of a spectrofluorometer differs from that of the spectrophotometer in two important respects besides other minor variations.

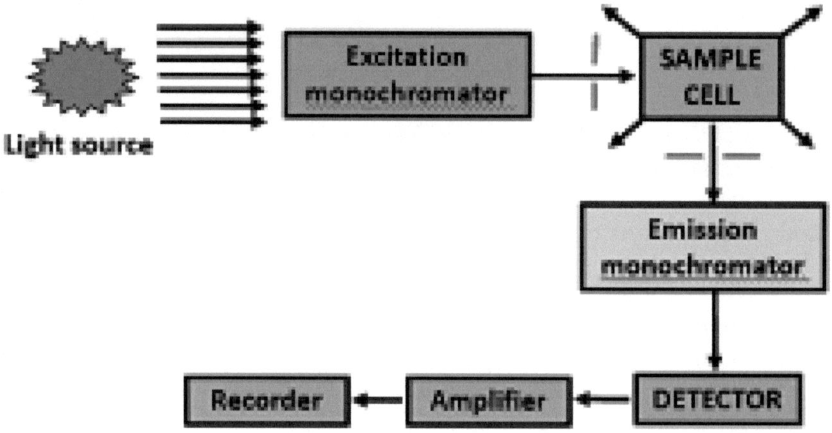

Figure 15.
Instrumentation of spectrofluorometer.

a. There are two monochromators instead of one as in a spectrophotometer; one monochromator is placed before the sample holder and one after it, and

b. As fluorescence is maximum between 25 and 30°C, the sample holder has a device to maintain the temperature.

The main components of a spectrofluorometer indicated in **Figure 15** are:

2.5.4 Applications

a. Identification of 3-D structure of protein: proteins are made by the combined cluster of 20 amino acids. For computational drug discovery it is very important to know that the correlation between three-dimensional structure and its functions. This data may not be successfully provided by the X-ray crystallography and electron microscopy. Hence, to resolve this problem the fluorescence spectrophotometry provides the reliable information about three-dimensional structure of protein by preserving the native structure of protein [35].

b. In food: fluorescence spectroscopy plays an important role in the determination of numerous food components, adulterants, additives, and contaminants [36].

c. Quality control in food processing: fluorescence spectroscopy is a rapid and sensitive analytical method for characterizing the food products. For example, recently, some authors have studied on impact of heat treatment on vitamin A in milk sample by using fluorescence spectra. The fluorescence spectra shows that the heat treatment induced a decrease in the fluorescence intensity at both 320 and 290 nm since milk samples heated at 75°C for 10 min. Compared with milk samples heated at 55°C during the same time. The details are shown in the (**Figure 16**) [37].

d. The more common applications of spectrofluorometer include qualitative analysis, quantitative analysis (applications include assay of riboflavin, thiamine, hormones such as cortisol, estrogens, serotonin and dopamine, organophosphorus pesticides, tobacco smoke carcinogens, drugs such as lysergic acid and barbiturates. Porphyrins. cholesterol, and even some metal ions); and studies on protein structure (FAD containing proteins).

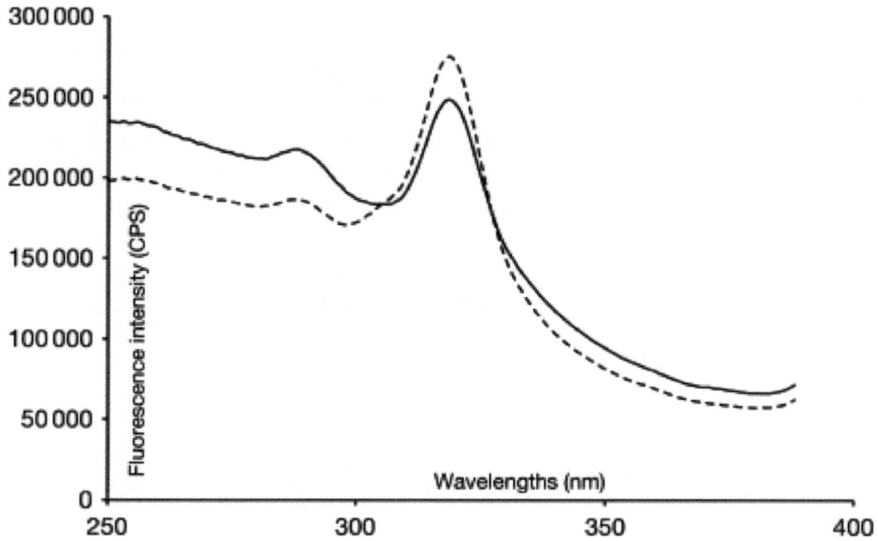

Figure 16.
Development of vitamin A fluorescence spectra acquired on milk samples after heating at 55°C and 75°C during 10 min.

2.6 Atomic absorption spectrophotometry (AAS)

2.6.1 Principle

When the sample molecules are subjected to volatilization, the produced atoms absorb certain wavelength of light from a source which produce a characteristic atomic spectrum of the molecule.

2.6.2 Instrumentation for AAS

The basic components of an atomic absorption spectrophotometer are [38]

a. Atomizers: most used atomizers in AAS are electrothermal atomizers (ETAs). These are involved to convert sample molecules into an individual gaseous atom that can absorb light from the source.

b. Light source: generally, hallow cathode lamp used as light source in AAS to produce a certain wavelength of light that are ideally absorbed by the gaseous atoms. Now a days, instruments with single and double beam optics are available.

c. Isolation and quantification of wavelengths of interest can be done by detectors and to control instrument operation and collect and process data the computer system helps.

2.6.3 Applications

a. AAS is a sensitive and highly selective spectrometric technique for the determination of many elements at trace and ultra-trace levels viz., Cd, Cr, Zn, Cu, Ag, Mn, Mg, Hg, As, Sc etc., at the picogram levels in soil, sediments and in plant samples. For example, this technique is very helpful for the scientists who are working on finding out the impacts of mining and industrial activities.

b. Atomic absorption spectrometry (AAS) can be used in the estimation of metals in body fluids like in serum, whole blood, apart from urine and in tissues for toxicological investigation in clinical studies [39].

c. AAS found to be a very good tool especially for the determination of food quality. In this concern, it is involved in the determination of alkaline and alkaline-earth metals because these elements act as a micro component in food samples.

d. To check the quality of water the AAS is one of the very useful techniques.

2.7 Nuclear magnetic resonance spectroscopy (NMR)

2.7.1 Principle

The nuclear magnetic resonance (NMR) is one of the most useful and powerful techniques in determination of molecular structure. The principle behind NMR is that when a strong magnetic field and a radiofrequency transmitter are applied on sample molecules, the atomic nuclei of those molecules get excited and forms spectral lines in the spectrum [40].

2.7.2 Instrumentation

The components of NMR spectrometer are

a. Radiation source: a radiofrequency transmitter (RF), which generates the radiofrequency current. This current deliver to the transmitting coil which creates a signal used to excite protons in the magnetic field.

b. A superconducting magnet, which produces magnetic field in the central volume of the magnet. The produced magnetic field varies from 1 to 10 T. The advanced NMR spectrometers consist superconducting solenoids that generates magnetic field above 3.5 T. The magnet has a bore to hold the sample probe and a room-temperature shim (RTS) coil assembly to reduce inhomogeneity of the magnetic field across the active sample volume.

c. A receiver receives absorbed signal in digitalized form generally called as free induction decay (FID).

d. A computer, an amplifier, and an analog-to-digital converter (ADC).

2.7.3 Applications

a. Biochemistry: NMR is a powerful technique in metabolic research. It became a method of choice for discovering the dynamics and compartmentation of metabolic pathways and networks [41]. Apart from this NMR is also useful in the study of intact biological specimens such as heart, kidney, and skeletal muscle with ^{31}P isotope.

b. Pharmacy: NMR spectroscopy permits the visualizing single atoms and molecules in various liquids as well as in solid state [42]. It is nondestructive and gives an idea on molecular structure that allows the structure elucidation and quantification of various organic molecules and provides information about the chemical structure and the dynamics of organic molecules in biological systems [35]. These structural studies provide an information related to functions of organic molecules such as amino acids, proteins, carbohydrates, and antibiotics such as ciprofloxacin, azithromycin and valinomycin.

c. Chemistry: NMR spectroscopy is used unambiguously to identify novel compounds, and as such, is usually required by scientific journals for identity confirmation of synthesized new compounds.

d. NMR has been used to study *in vivo* or synthetic membrane transport system. For example, it has been used for study of transport of Na^+ ions in human RBCs.

e. NMR has been used for the quantitative determination of the concentration of metabolites. Using NMR in the determination of concentration of phosphocreatine in human muscle is an example of the same [7].

f. NMR is a very useful techniques in identifying and quantifying hydrocarbons in petroleum industry. For example, the ^1H and ^{13}C nuclear magnetic resonance is involved in the quantitative measurement of liquid hydrocarbons in FACE Gasoline F. It gives a good resolution in the spectrum which is useful for quantifying liquid hydrocarbons in FACE Gasoline F [43].

2.8 Electron spin resonance spectroscopy (ESR)

Electron spin resonance (ESR) or electron paramagnetic resonance (EPR) spectroscopy is an analytical technique for detecting and characterizing the paramagnetic species.

2.8.1 Instrumentation

Figure 17 illustrates the basic components of an ESR spectrometer. Fields of 50–500 mT, required for accurate work are generated by electromagnets. Auxiliary sweep generators with a capacity of 10–100 mT are also provided. Monochromatic microwave radiation might be readily obtained by using a klystron oscillator.

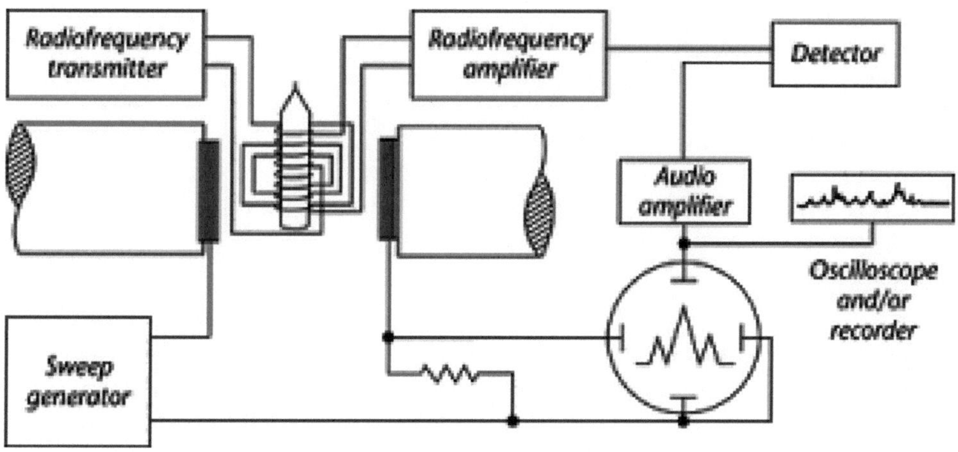

Figure 17.
Components of NMR spectrometer.

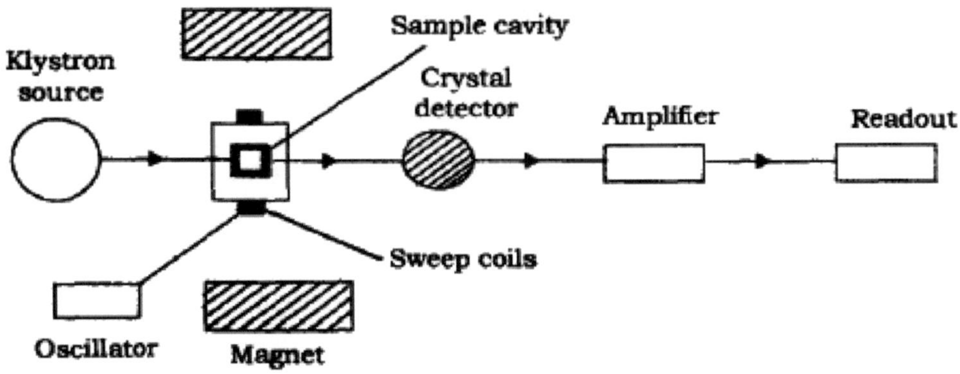

Figure 18.
Basic components of an electron spin resonance spectroscopy.

Samples for ESR must be solids. Biological samples which contain a large amount of water are therefore frozen in liquid nitrogen before ESR experiment. **Figure 18** shows the Basic components of an ESR [7].

2.8.2 Principle

The principle of ESR spectroscopy is based on the fundamental properties of electrons. An unpaired electron in atoms or molecules possesses paramagnetic character and shows both magnetic moment and angular momentum. In an external magnetic field, the spin magnetic moment aligns parallel or antiparallel to the field, and the spinning electrons are split or divided into high and low energy states. These split electrons form spectral lines, and they can be determined [44].

2.8.3 Applications

a. Electron spin resonance spectrometry is used in polymer chemistry to identify paramagnetic species in polymers. During polymer synthesis or degradation, the paramagnetic species are produced due to free radical mechanism. These paramagnetic species may damage the quality of polymer. Hence, ESR provides an information about the number of paramagnetic species [45].

b. ESR has been used in studying reaction mechanisms which proceed through free radical intermediates in metabolic reactions.

c. ESR is one of the important methods to study transition metals present in biological systems such as iron in hemoglobin, and cytochromes, copper in cytochrome oxidase, molybdenum in xanthine oxidase etc.

3. Conclusion

Spectroscopy is one of the most important analytical tools for the analysis of various compounds in various fields including chemistry, physics, biology, agriculture, engineering, and medicine. This is working with different principles which are projected through various instrumentation techniques like UV-visible spectrophotometry, IR spectroscopy, Raman spectroscopy, NMR spectroscopy and ESR spectroscopy. The applications of these techniques are providing very useful information to the teachers, students, and researchers. These analytical methods are nondestructive, consistent, and reliable and require no or very little sample preparation. We can apply these methods to solid, liquid, and powdered samples. They can be used for a wide range of elemental analysis and provide detection limits at the sub-ppm level concentrations easily and simultaneously. Therefore, the spectrophotometric techniques are very useful in elementary analysis in all most all fields by providing reliable information about elements to be analyzed.

Author details

Murali Dadi[1*] and Mohd Yasir[2]

1 Department of Chemistry, School of Mathematics and Natural Sciences, Copperbelt University, Kitwe, Zambia

2 Department of Pharmacy, School of Health and Medical Sciences, Arsi University, Asella, Ethiopia

*Address all correspondence to: murali.dadi@gmail.com; murali.dadi@cbu.ac.zm

References

[1] Germer TA, Zwinkels JC, Tsai BK. Theoretical concepts in spectrophotometric measurements. Experimental Methods in the Physical Sciences. 2014;**46**:11-66

[2] Trumbo TA, Schultz E, Borland MG, Pugh ME. Applied spectrophotometry: Analysis of a biochemical mixture. Biochemistry and Molecular Biology Education (Wiley Online Library). 2013;**41**(4):242-250

[3] Rand RN. The role of spectrophotometric standards chemistry laboratory. Journal of Research of the Notional Bureau of Standards-A. Physics and Chemistry. 1972;**76**(5):499-508

[4] Zwinkels J. In: Luo R, editor. Light, Electromagnetic Spectrum. Encyclopedia of Color Science and Technology. New York: Springer Science and Business Media; 2015. DOI: 10.1007/978-3-642-27851-8_204-1

[5] https://chem.libretexts.org/Bookshelves/Physicaland TheoreticalChemistryTextbook

[6] Kruschwitz JDT. Chapter 61, Spectrophotometer Components. Field Guide to Colorimetry and Fundamental Color Modeling. 2018. DOI: 10.1117/3.2500912.ch61

[7] Upadhyay A, Upadhyay K, Nath N. Biophysical Chemistry (Principals and Techniques). Girgaon, Mumbai, India: Himalaya Pub House; 2009. ISBN: 1282802003 9781282802001

[8] De Caro CA, Claudia H. UV/VIS Spectrophotometry—Fundamentals and Applications. Schwerzenbach, Switzerland: Mettler-Toledo Publication No. ME-30256131; 2015

[9] Murali D, Venkatrao SV, Rambabu C. Spectrophotometric determination of etravirine in bulk and pharmaceutical formulations. American Journal of Analytical Chemistry. 2014;**5**:77-82

[10] Verevkin SP, Zaitsau DH, Schick C, Heym F. Development of direct and indirect methods for the determination of vaporization enthalpies of extremely low-volatile compounds. Handbook of Thermal Analysis and Colorimetry. 2018;**6**:1-46. DOI: 10.1016/B978-0-444-64062-8.00015-2

[11] Ling Li X, Hu YJ, Mi R, Yun Li X, Qi Li P, Ouyang Y. Spectroscopic exploring the affinities, characteristics, and mode of binding interaction of curcumin with DNA. Molecular Biology Reports. 2013;**40**:4405-4413

[12] Passos MLC, Sarraguça MC, MLMFS S, Rao TP, Biju VM. Spectrophotometry | organic compounds. In: Reference Module in Chemistry, Molecular Sciences and Chemical Engineering. Encyclopedia of Analytical Science (Third Edition). Amsterdam, Netherlands: Elsevier Ltd; 2019. pp. 236-243. DOI: 10.1016/b978-0-12-409547-2.14465-8

[13] Rojas FS, CanoPavón JM. Spectrophotometry | Biochemical Applications. Encyclopedia of Analytical Science. 2nd ed. Amsterdam, Netherlands: Elsevier/The Lancet publishers, Elsevier Ltd; 2005. pp. 366-72

[14] Leong YS, Ker PJ, Jamaludin MZ, Nomanbhay SM, Ismail A, Abdullah F, et al. UV-vis spectroscopy: A new approach for assessing the color index of transformer insulating oil. Sensors (Basel). 2018;**18**(7):2175

[15] Yong Wang YN. Combination of UV–vis spectroscopy and chemometrics to understand protein–nanomaterial conjugate: A case study on human serum albumin and gold nanoparticles. Talanta. 2014;**119**:320-330

[16] Ahmed S, Mahesar LM, Durazzo A, Santini A, Anja I, et al. Application of infrared spectroscopy for functional compounds evaluation in olive oil: A current snapshot. State-of-the-art infrared applications in drugs. Dietary Supplements, and Nutraceuticals. 2019;**2019**:1-11. DOI: 10.1155/2019/5319024

[17] Colthupa NB. Infrared Spectroscopy. Encyclopedia of Physical Science and Technology. 3rd ed. Bridgewater Township, New Jerse: American Cyanamid Company; 2003. pp. 793-816. DOI: 10.1016/B0-12-227410-5/00340-9

[18] Simek JW, Wade LG. Organic Chemistry. 7th ed. New York, USA: Pearson Education, Inc; 2010

[19] Yaneva Z, Georgieva N. Chapter-5, physicochemical and morphological characterization of pharmaceutical nanocarriers and mathematical modeling of drug encapsulation/release mass transfer processes. In: Nanoscale Fabrication, Optimization, Scale-Up and Biological Aspects of Pharmaceutical Nanotechnology. Norwich, NY: William Andrew Applied Science Publishers; 2018. pp. 173-218. DOI: 10.1016/b978-0-12-813629-4.00005-x

[20] Hamers RJ, Franking XWR, Ruther R, Stavis C. Chapter 3—Infrared spectroscopy for characterization of biomolecular interfaces. In: Biointerface Characterization by Advanced IR Spectroscopy. Elsevier B.V.; 2011. pp. 57-82. DOI: 10.1016/B978-0-444-53558-0.00003-5

[21] Kumar S, Barth A. Following enzyme activity with infrared spectroscopy. Sensors (Basel). 2010;**10**(4):2626-2637

[22] Kereszturi A, Gyollai I, Szabó M. Case study of chondrule alteration with IR spectroscopy in NWA 2086 CV3 meteorite. Planetary and Space Science. 2015;**106**:122-131

[23] Titus D, Samuel EJJ, Roopan SM. Chapter 12—Nanoparticle characterization techniques. In: Green Synthesis, Characterization and Applications of Nanoparticles. A Volume in Micro and Nano Technologies. 1st ed. 2019. pp. 303-319. DOI: 10.1016/b978-0-08-102579-6.00012-5

[24] Mohamed Shameer P, Mohamed NP. Chapter-8, Exploration and enhancement on fuel stability of biodiesel: A step forward in the track of global commercialization. In: Advanced Biofuels. Applications, Technologies, and Environmental Sustainability. Lambert Academic Publishing; 2019. pp. 181-213

[25] Sindhu R, Binod P, Pandey A. Chapter-17, microbial poly-3-hydroxybutyrate and related copolymers. In: Industrial Biorefineries& White Biotechnology. 1st ed. Elsevier B.V; 2015. pp. 575-605. DOI: 10.1016/b978-0-444-63453-5.00019-7

[26] Kowalczuk D, Pitucha M. Application of FTIR method for the assessment of immobilization of active substances in the matrix of biomedical materials. Materials. 2019;**12**(18):1-13

[27] Zhang Ji, Li B, Wang Q, Wei X, Feng W, Chen Y, et al. Application of Fourier transform infrared spectroscopy with chemometrics on postmortem interval estimation based on pericardial fluids. Scientific Reports. 2017; 7(18013):1-8

[28] Bumbrah GS, Sharma RM. Raman spectroscopy—Basic principle, instrumentation, and selected applications for the characterization of drugs of abuse. Egyptian Journal of Forensic Sciences. 2016;**6**:209-215

[29] https://www.slideshare.net/aravi_ab/raman-spectroscopy-34530148

[30] Kalantri PP, Somani RR, Makhija DT. Raman spectroscopy: A

potential technique in analysis of pharmaceuticals. Der Chemica Sinica. 2010;**1**(1):1-12

[31] Rangan S, Schulze HG, Vardaki MZ, Blades MW, Piret LM, Turner RFB. Applications of Raman spectroscopy in the development of cell therapies: State of the art and future perspectives. Analyst. 2020;**145**:2070-2105

[32] Ramos IRM, Malkin A, Lyng FM. Current advances in the application of Raman spectroscopy for molecular diagnosis of cervical cancer. BioMed Research International. 2015;**2015**:1-9

[33] Weng S, Zhub W, Zhang X, Yuan H, Zheng L, Zhao J, et al. Recent advances in Raman technology with applications in agriculture, food and biosystems: A review. Artificial Intelligence in Agriculture. 2019;**3**:1-10

[34] Buckley K, Ryder AG. Applications of Raman spectroscopy in biopharmaceutical manufacturing: A short review. Applied Spectroscopy. 2017;**71**(6):1085-1116

[35] Misra G. Fluorescence Spectroscopy. Data Processing Handbook for Complex Biological Data Sources. 1st ed. Academic Press; 2019. pp. 31-37

[36] Gómez-Hens A. Fluorescence | food applications. In: Encyclopedia of Analytical Science. 2nd ed. Amsterdam, Netherlands: Elsevier Ltd; 2005. pp. 186-194. DOI: 10.1016/b0-12-369397-7/00172-2

[37] Karoui R. Quality Control in Food Processing. Reference Module in Food Science, Encyclopedia of Food and Health. Amsterdam, Netherlands: Elsevier Ltd; 2016. pp. 567-572. DOI: 10.1016/b978-0-12-384947-2.00582-1

[38] Butcher DJ. Atomic Absorption Spectrometry | Interferences and Background Correction. Encyclopedia of Analytical Science. 2nd ed. Amsterdam, Netherlands: Western Carolina University, Cullowhee, NC, USA, Elsevier Ltd; 2005. pp. 157-163. DOI: 10.1016/b0-12-369397-7/00025-x

[39] Calatayud JM, Icardo MC. Flow Injection Analysis, Clinical and Pharmaceutical Applications. Encyclopedia of Analytical Science. 2nd ed. Amsterdam, Netherlands: Elsevier Ltd; 2005. pp. 76-89. DOI: 10.1016/b0-12-369397-7/00159-x

[40] Günther H. NMR Spectroscopy: Basic Principles, Concepts and Applications in Chemistry. 3rd ed. John Wiley & Sons, Ltd., Chichester, U.K; 2013. p. 734

[41] Fan TWM, Lane AN. Applications of NMR spectroscopy to systems biochemistry. Progress in Nuclear Magnetic Resonance Spectroscopy. 2016;**92-93**:18-53

[42] Diehl B. Principles in NMR Spectroscopy. NMR Spectroscopy in Pharmaceutical Analysis. 1st ed. Elsevier Ltd.; 2008. pp. 3-41

[43] Ure AD, O'Brien JE, Dooley S. Quantitative NMR spectroscopy for the analysis of fuels: A case study of FACE Gasoline F. Energy Fuels. 2019;**33**(11):11741-11756

[44] Shah JM. ESR Insights into Macroradicals in UHMWPE. UHMWPE Biomaterials Handbook. In: Ultra-High Molecular Weight Polyethylene in Total Joint Replacement and Medical Devices. 3rd ed. Norwich, NY: William Andrew Applied Science Publishers; 2016. pp. 433-450

[45] Schlick S, Jeschke G. Electron spin resonance spectroscopy. Polymer Science: A Comprehensive Reference. 2012;**2**:221-253

Section 3

Colour Measurement and Application for Textile and Food Industry

Chapter 5

Basic Principles of Colour Measurement and Colour Matching of Textiles and Apparels

Pubalina Samanta

Abstract

This book chapter covers basic principles of quantitative measurement and analysis of surface colour parameters and surface appearance of undyed/dyed textile materials and finally matching of colours with standard samples of any textiles. Surface colour parameters of textile materials change with different chemical processing including bleaching, dyeing and finishing and need to measure it quantitatively for understanding the effect of different chemical processes/dyes and auxiliaries and finishes. So, CIE-1976 equations for measurement of Tristimulus values, surface colour strength, colour differences, Metamerism index, colour difference index as well as specific formulae for measuring Whiteness Index, Yellowness Index, Brightness Index and the theory of colour match prediction are discussed here. One Case study of colour match prediction for a specific case is also shown. Finally, the importance of single constant measurement of surface color parameters for coloured textiles and practical cares for database preparation and colour measurement and match prediction for textile and apparel products are deliberated. Apparel industry is very much dependent on colour psychology and colour preferences of customers in different seasons and occasions and hence, it is important to measure all surface colour parameters of textile materials to choose perfectly matched coloured textiles for making any garment.

Keywords: surface colour strength, Tristimulus values, Total colour differences, colour difference index, Metamerism, whiteness-yellowness and brightness indices. Computer aided colour matching

1. Introduction

Colour and aesthetics are as important as its various physical properties for textiles/garments, leather, moulded plastics,and products of various other fields. The ability to integrally colour(dyeing and Printing) textiles, leather, plastic moulded articles has an important edge over others non polymeric rigid hard materials like metals etc.

Matching of colours and imitate texture, especially in specific textiles of different fibres and blends is crucial in many of its apparel applications. The task becomes more difficult when colours need to be exactly matched with standard colour yield desired with endurable criteria for acceptable colour fastness to wash, rubbing, light and perspiration etc. for different textiles/plastics and polymer products. So, understanding Theory of colour measurement, quantification, for well defined applications of different dyes/pigments on different textiles materials has become a must for the textile or leather dyers/printers and plastics injection moulder/wall paints etc.

Day by day, more and more concern of consumers on colour matching for consumers' textiles and live-style products of apparels and furnishings, bed linen and auto-mobile (car) Interiors, appliances and along with polymer/plastic assemblies, are pushing the dyed/printed coloured textile product manufacturer to develop their products with more precision colour matching with least meta-merism.

In order to understand the colour we have to know, how the colour is perceived. The perception of colours [1] involves the interaction of three elements. (i) source of light, (ii) an object and (iii) human eye.

2. Colour theory

Colour can be broadly defined as the physio-psychometric effect on the brain of an observer from reflected wavelength of an object, when that object is viewed in presence of a definite light source.

Colour theory meant a Standardised scientific method with specific mathematical/empirical formula with arrangement of incidence of light/standard illuminant for absorption and reflection of the colour on and from the object and then detecting folowed by measurement of colour value specific reflactance or any other quantified values to record and communicate colour information for reproducibility and matching.

2.1 CIE definition 845-02-18 of perceived colour

As per CIE definition 845-02-18: (perceived) co***lour*** may be defined [2] or perceived as " Attribute of a visual perception consisting of any combination of chromatic and achromatic content. This attribute can be described by chromatic color names such as yellow, orange, brown, red, pink, green, blue, purple, etc., or by achromatic color names such as white, gray, black, etc., and qualified by bright, dim, light, dark etc., or by combinations of such names [Unquote].

Human Perception of colour usually appear to describe a colour in terms of amount of RGB (Red, Green and Blue) sensation of human eye as an additive colour mixing system (as shown in **Figure 1**) distinguish among qualitative and geometric differences of colour perceptions by its predominating Hue (predominating Reflected wavelength observed), Value (Light or Dark i.e. White or Black

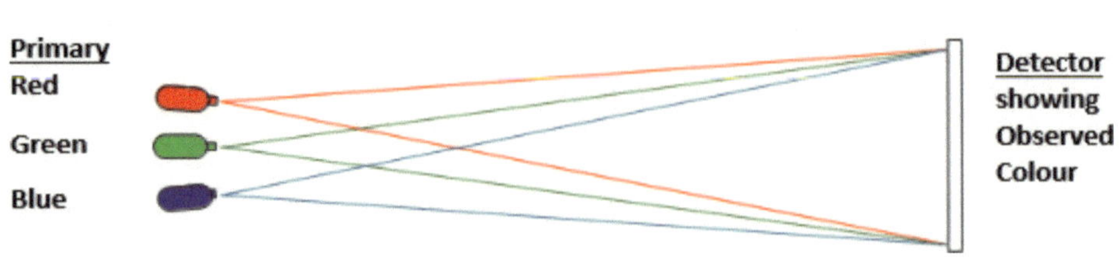

Figure 1.
Schematic display of colour observed from RGB primary colour stimuli with standard illuminant by a standard observer /detector.

respectively) and chroma (Strength or Concentrations of Coloured mol.) [1] with/without associated brightness/dullness and uv absorption.

Thus, Colour of any object can be considered as an physio-psychological illusion of reflected radiation / visible light from a substance/object after incidence of light on it, which is to be detected in exact quantitative terms by amount of RGB in it (**Figure 1**), while, colorimetry is the measurement and evaluation technique of colour value in any quantifiable terms by which this physio-psychological sensation of our eyes can be converted to the actual physical measurement of colour values either in solution or in solids in some sensory values of primary colours.

If any one want to buy a skirt or a pair of slacks to match a jacket, one cannot match the colour by memory — he/she has to take the jacket with him/her to match it visually by judging colour by eye measurement in the store itself. it may happen that in some cases, the store light is insufficient and faulty matching by eyes results, So, one has to match it also under standard and sufficient incandescent light in the dressing room and and also in the outdoor sunlight too.

Three fundamental components of understanding or measuring or matching any colour:

- light sources /standard illuminant
- objects / samples illuminated by standard light
- observers /detector to record colours reflected from it

Most Important and commercially useable Two of the major Colour quantification systems are:

- Munsell Colour Theory -hue, value and chroma
- The CIE Theory of Colour -Tristimulus values (X, Y and Z) and CIE L*, a*and b* values

The CIE Theory of Colour −1931 updated in 1976 [1.2] is in wide commercial use for textile's colour communication and hence it has significant importance in apparel sector and is described below.

2.2 CIE 1931 standard theory of colour

The Commission Internationale de l'Éclairag e (CIE) has recommended x(l), y(l), z(l) for lÎ [360 nm, 830 nm] in 1 nm steps for distinguishing colour based on tristimulas values, which are well accepted/.

Over a wide range of conditions of observation, many colour stimuli can be matched in colour completely by additive mixtures of three fixed primary colour stimuli (RGB) combine linearly, symmetrically, and transitively and may be expressed as X (stimuli of Red Primary), Y (stimuli of Green primary), and Z (*Stimuli of Blue Primary) as three coordinates called Tristimulus values (X, Y and Z) or stimuli of any colour observed under standard illuminant under standard detector / observer, based on reflected contributions of R G B primaries from any object (**Figure 1**) having spectral sensitivity of RGB primary stimuli (as shown in **Figure 2a**) and actual Tristimulus values of any object (X, Y and Z values, as per formulation shown in **Figure 2b**).

Thus, a monochromatic colour stimulus ($\mathbf{Q_l}$) of wavelength L, it can be represented /expressed as

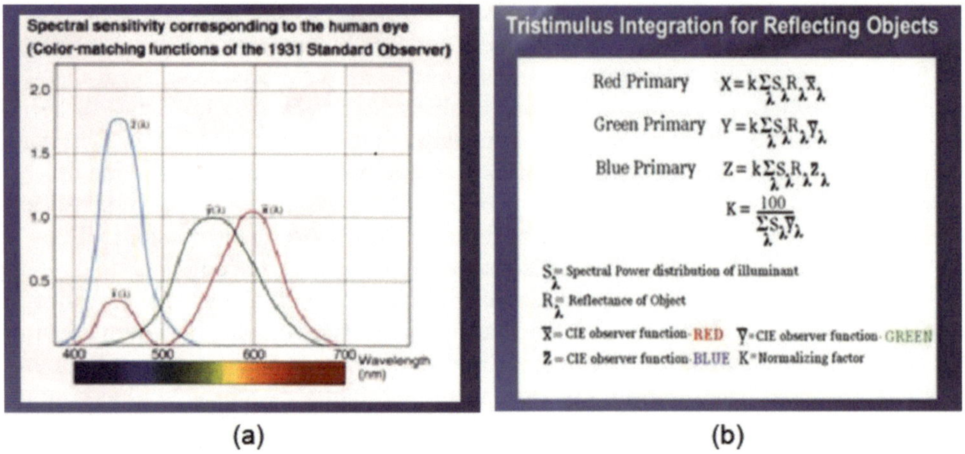

Figure 2.
(a): Spectral sensitivity of human eyes of RGB primary Stimulli and (b): Resultant actual measurement of reflectance values and corresponding Tristimulus values of any object.

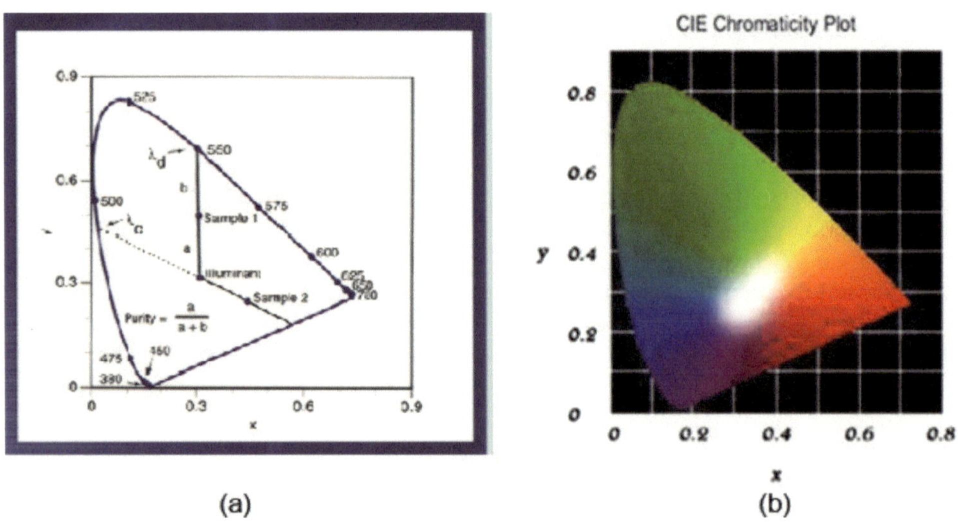

Figure 3.
a: CIE 1931 chromaticity diagram showing b: CIE 1931 chromaticity diagram all predominating visible hues at 380-700 nm (coloured) showing x and y coordinates.

$$Q_l = X_l\boldsymbol{R} + Y_l\boldsymbol{G} + R_l\boldsymbol{B},$$

where R_l, G_l, and B_l are the spectral *tristimulus values* of $\mathbf{Q_l}$.

Consequently, the **CIE 1931 chromaticity diagram** [1–3] is **not** a perceptually uniform chromaticity space from which the perception of chromaticity can be derived, where (**Figure 3a** and **b**)

$$\boldsymbol{x} = \boldsymbol{X}/(\boldsymbol{X} + \boldsymbol{Y} + \boldsymbol{Z}), \quad (1)$$

$$\boldsymbol{y} = \boldsymbol{Y}/(\boldsymbol{X} + \boldsymbol{Y} + \boldsymbol{Z}), \quad (2)$$

$$\boldsymbol{z} = \boldsymbol{Z}/(\boldsymbol{X} + \boldsymbol{Y} + \boldsymbol{Z}), \quad (3)$$

$$\textit{and}\ \boldsymbol{x} + \boldsymbol{y} + \boldsymbol{z} = \mathbf{1} \quad (4)$$

and also $\boldsymbol{z} = [\mathbf{1}\text{-}\boldsymbol{x}\text{-}\boldsymbol{y}]$.

Hence there is no need to plot a three dimensial x,y and z diagram, rather the 2 dimensional chromaticity coordinate plot (**Figure 3a** or **b**) is sufficient to get x,y & z values from x vs. y plot of 2 dimensional CIE chromaticity diagram [1–3].

As per 1976 CIE L*a*b* colour space, L*, a* and b* values for comparing colour differences in betrween two samples of same or similar textile fabrics n be represented by ΔL^***,** Δa^* **and** Δb^* values [1–3].

CIE 1976 lightness/darkness is represented by, L^* **(L = 0 black and L-100 white)**, more or less similar to lightness distribution to the Munsell Value scale and, CIE 1976 scale of **redness** ($a^{*\,-ve}$) **/greenness** (a^{*-ve}) is represented by a^* and **CIE 1976 scale of yellowness** (b^{*+ve}) **/blueness** ($b^{*\,-ve}$) is represented by b^**, and* **1976 CIE L*a*b* colour space diagram** represent total colour differences value by ΔE *or* ΔE * **(as shown in Figures 4 and 5).**

$$\text{Where, } \mathbf{L^* = 116(Y/Y_o)^{1/3}–16}, \quad \Delta L^* = L^*{}_1–L^*{}_2 \tag{5}$$

$$\mathbf{a^* = 500\left[(X/X_o)^{1/3}–(Y/Y_o)^{1/3}\right]}, \quad \Delta\mathbf{a^* = a^*{}_1–a^*{}_2} \tag{6}$$

$$\mathbf{b^* = 200\left[(Y/Y_o)^{1/3}–(Z/Z_o)^{1/3}\right]}, \quad \Delta\mathbf{b^* = b^*{}_1–b^*{}_2} \tag{7}$$

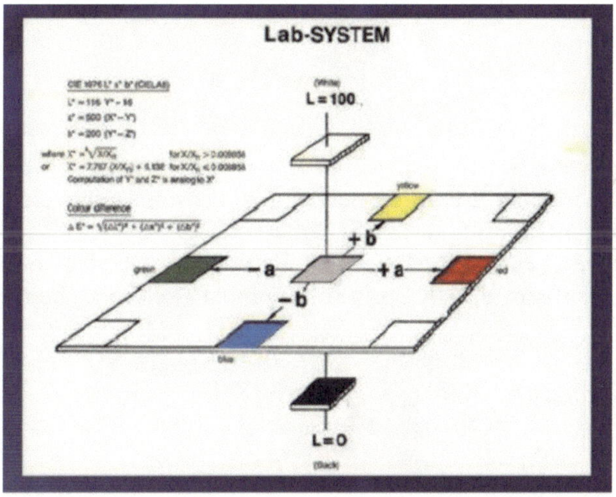

Figure 4.
CIE-L a* b* system of colour difference plot to determine ΔL*, Δa* and Δb* values of a pair.*

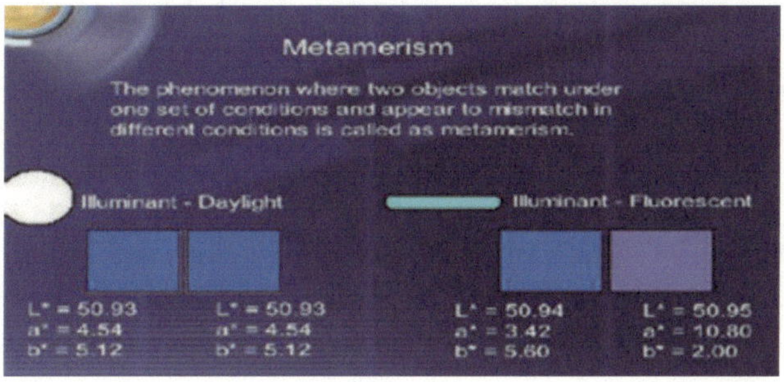

Figure 5.
Example of metameric match under two different illuminant (day light and Fluoroscent light conditions.

Chroma, (psychometric chroma) values in CIELAB colour space [1–3] was calculated as follows:

$$\mathbf{C}_{(\mathbf{ab})^*} = \left(\mathbf{a}^{*2} + \mathbf{b}^{*2}\right)^{1/2}, \quad \Delta\mathbf{C}^* = \mathbf{C}^*{}_{1(\mathbf{ab})} - \mathbf{C}^*{}_{2(\mathbf{ab})} \tag{8}$$

Where, $C^*_{1(ab)}$ and $C^*_{2(ab)}$ are the chroma values for standard and produced sample.

CIE 1976 metric Hue-Difference (ΔH) for CIELAB system [1–3] was calculated as follows:

$$\Delta\mathbf{H}_{\mathbf{ab}} = \left[(\Delta\mathbf{E}_{\mathbf{ab}}{}^*)^2 - (\Delta\mathbf{L}^*)^2 - \left(\mathbf{C}_{\mathbf{ab}}{}^*)^2\right)\right]^{1/2} \tag{9}$$

Total Colour Differences [1–3]:

$$\Delta\boldsymbol{E}^* = \left[(\Delta\boldsymbol{L}^*)^2 + (\Delta\boldsymbol{a}^*)^2 + (\Delta\boldsymbol{b}^*)^2\right]^{½} \tag{10}$$

As per Kubelka Munk Equation [1–3]. Surface colour Strength (K/S value) of Coloured flat surface is:

$$\begin{aligned} \mathrm{K/S} &= \frac{\text{Co-efficient of absorption}}{\text{Co-efficient of scattering}} \\ &= \frac{(1 - \mathrm{R}_{\lambda\,\max})^2}{2\mathrm{R}_{\lambda\,\max}} = \alpha \mathrm{C_D} \end{aligned} \tag{11}$$

Where K is the coefficient of absorption; S, the coefficient of scattering; and $R_{\acute{\lambda}max,}$ is the Reflectance value at maximum absorbance wavelength (λmax) and C_D is the dye concentration and α is the constant.

For use mixture of colourants to obtain a compound shades on the surface of textiles or similar substrate, as K/S is additive in nature, it can be represented as follows:

$$\begin{aligned} (\mathrm{K/S})_{\mathrm{MAX}} &= (\mathrm{K/S})_{\mathrm{subs}+}(\mathrm{K/S})_1 + (\mathrm{K/S})_2 + (\mathrm{K/S})_3 + \dots \dots \\ &= (\mathrm{K/S})_{\mathrm{subs}} + \lambda_1 \mathrm{C}_1 + \lambda_2 \mathrm{C}_2 + \lambda_3 \mathrm{C}_3 + \dots \dots . \end{aligned} \tag{12}$$

Moreover, plot of dye concentration Vs K/S value is linear in relation and is easier to predict for any unknown concentration of dyes, while plot of Reflectance vs. dye concentration is non linear in relation and is difficult to predict and hence K/S is an important surface colour measuring criteria of textiles, against application of increase or decrease in dye concentration on same or similar fabrics.

For coloured textile materials, it may be presumed that dye molecules are not contributing to change in scattering and therefore K integrated over visible wavelength is the total sum of absorption of dyestuff and textile substrate (so, if substrate is not changed, scattering is not changed). Therefore, surface colour strength i.e. K/S value is directly proportional with concentration of dye molecules and the scattering value of the dyed textile sample is not dependent on the concentration of dye stuff (but that is not true for the case of pigments in paint or binded over textile substrate with binder chemicald and fixed on the surface of the materials). So, for textile substrate, it is single constant theory of Colour Matching applicable for all textile materils.

Hence, for any textile surface, for the particular dyed/coloured textile sample (Fibre, Yarn and Fabric construction and type and amount of surface deposited (coated or impregnated)finish remains un-altered), and scattering values remains constant always, if textile fabric used is not changed.

Thus higher is the K/S value, meant higher is the absorption value, meant higher absorption value signifying or indicating higher surface dye uptake, governed by following formulae.

$$(K/S)_{mix} = C_1(K/S)_1 + C_2(K/S)_2 + C_3(K/S)_3 \tag{13}$$

$$(K/S)_{dyed\ substrate} = C_0(K/S)_{un\ dyed\ substrate} + C_1(K/S)\ d_1 + C_2(K/S)d_2 + C_3(K/S)d_3 \tag{14}$$

Finally, reflectance Vs. dye concentration is not linear & is difficult to interpolate or curve fitting.

While K/S Vs. dye concentration is linear can be interpolated thus can be used in computerised colour measurement and matching software·

2.3 Principles of colour matching of dyed textiles

A colour match between two sets of samples means:

Colour of produced sample = Colour of given standard sample i.e. (X_{SL},Y_{SL},Z_{SL}) values of produced sample = (X_{SD},Y_{SD},Z_{SD}) values of given standard sample while X,Y & Z are the tristimulus value of produced Sample (SL) and Standard (SD) sample Also match may be predicted or judged [1–3] by $(Reflectance)_{SL}$ of produced sample integrated at 400 to700 nm = $(Reflectance)_{SD}$ of standard sample integrated at 400 to 700 nm or $(K/S)_{SL}$ values of produced sample = $(K/S)_{SD}$ value of standard sample, where K/S = α C_D and K/S values (as per kubelka munk Equation [1–3]) as stated above), is:

$$\begin{aligned} K/S &= \frac{\text{Co-efficient of absorption}}{\text{Co-efficient of scattering}} \\ &= \frac{(1 - R_{\lambda\,max})^2}{2R_{\lambda\,max}} = \alpha C_D \end{aligned} \tag{15}$$

For a ternary mixture of colourants/dyes to obtain any particular compound shade on textiles, three equations are to be solved as a function of dye concentrations of the colourants (1,2,3 or n) and have to determine tristimulus values or K/S Values obtained through reflectance measurement of samples to match. More over K/S value being additive and dye concentration vs. K/S being linear in nature, the resultant K/S value of a dyed sample (dyed with mixture of three different dyes (d_1, d_2 and d_3) in respective concentrations (c_1, c_2 and c_3) is represented by following matrix equations:

$$\boxed{\begin{aligned} \mathbf{f_x(c_1, c_2, c_3) = X} \\ \mathbf{f_y(c_1, c_2, c_3) = Y} \\ \mathbf{f_z(c_1, c_2, c_3) = Z} \end{aligned}} \quad \text{and} \quad \boxed{\begin{aligned} &(K/S)_{mix} = c_1(K/S)d_1 + c_2(K/S)d_2 + c_3(K/S)d_3 \\ &(K/S)_{dyed\ substrate} = c_0(K/S)_{un\ dyed\ substrate} + c_1(K/S)d_1 \\ &+c_2(K/S)d_2 + c_3(K/S)d_3 \end{aligned}} \tag{16}$$

Where, **X, Y** and **Z** are the tri-colorimetric tristimulus values of samples produced i.e. to be matched finally and c_1,c_2, c_3 are the exact concentration of amount of dyes/colourants required.

In practical ceases, the reflectance values of given standard sample at 400 to 700 nm are determined from given standard coloured solid textile fabric surface and the obtained results of reflectance values at different wave length in visible region are processed in computer aided colour matching software for matching **X, Y and Z** values of produced samples or simply by comaparing the total integrated K/S values at 400–700 nm wavelength (visible range) for ultimate checking of colour matching of textiles in terms of allowable limits (tolerences) of colour difference values of **Δ L*, Δa* and Δb* values** representing **1976 CIE L*a*b* colour space diagram** and in terms of allowable limit of **total colour differences** value represented by ***ΔE / ΔE *.***

A textile match however should be ideally be an isomeric match i.e. match under all illuminant. But in actual practice, when two coloured sample show a match of colour under one illuminant may not match under other illuminantand this difference of match under specific two conditions of different illuminant is termed as illuminant based metameric match. Besides variation of illumainnat, there are other different types of metamerism for change of conditions of colour measurement, as follows, arised during colour matching under varying ambience of any one factor or others [1–3]:

Types of metamerism:

- **Illuminant metamerism**: example: daylight and D65 simulation fluorescent lamp
- **Object metamerism**: example: metameric inks (see metamerism kit)
- **Observer or Sensor metamerism**: example: Instrumental scanner and human visual perception
- **Complex metamerism and Instrument metamerism:** example:two inks/ dyed clothes measured under two different instruments.

Colour matching is therefore based on to find a Least metameric match where, General metamerism index [1–3] is as given below:

$$\text{General metamerism Index} = \frac{\sum(\Delta R\bar{x})^2}{X^2} + \frac{\sum(\Delta R\bar{y})^2}{Y^2} + \frac{\sum(\Delta R\bar{z})^2}{Z^2} \tag{17}$$

Where ΔR = Difference in reflectance between pair of metamer samples; $\bar{x}, \bar{y}, \bar{z}$ = CIE standard observer colour function X, Y, Z = CIE tristimulas value normally taken for illuminant C.

The Metamerism-Index (MI) shows the probability that two samples will show the same colour difference under two different illuminants (represented by the first and second illuminant) or under two different instruments or under any two different conditions of colour measurements. CIE LAB i.e. LABD metamerism index [1–3] is represented as:

$$MI_{(LABD)} = \left[(\Delta L_1{}^* - \Delta L_2{}^*)^2 + (\Delta a_1{}^* - (\Delta a_2{}^*)^2 + (\Delta b_1{}^* - (\Delta b_2{}^*)^2\right]^{1/2} \tag{18}$$

$\Delta L_1{}^*$, $\Delta a_1{}^*$, and $\Delta b_1{}^*$are the Delta CIELab* colour coordinates between Standard and Sample for the first illuminant and $\Delta L_2{}^*$, $\Delta a_2{}^*$, and $\Delta b_2{}^*$are the Delta CIELab* colour coordinates between Standard and Sample for the second illuminant Interpretation:

When, MI (metamerism index) is low, the colour difference between the sample pair (standard vs. produced) is the more closer and similar, for different conditions of measurement even under different illuminants.

Reflectance Spectrophotometers are very effective instrument in measuring and recording colours of any solid substrate / textiles from its substrate. All measurements are done by CIE −1976 System [1–3], which is commonly used in textile and other colour related industry.

The accuracy of colour matching of textiles depend on the set of tolerance limits [1–3] for **Δ L*, Δa*, Δb* and ΔE / ΔE *. values under pre decided illuminant as standard illuminant, which** is another significant task. It is generally thumb rule from practical observations that DE values below or within 1.0 are acceptable match i.e. DE values above 1is not a good match at all. Moreover, during dyeing in industry, all dyers and other relevant people likes to create more stringent tolerance values for colour matching of textile substrate preferably DE value are to be within 0.7 only. Setting of colour matching tolerences in terms of Colour differences values are discussed later in item 3.1 in details.

Thus these colour difference data has many benefits to the dyers entailing how much it is darker or light, how much it is redder or greener or how much it is bluer and yellower, when the shades of two nearly match samples are compared. Thus, Human eye estimated perception of colour differences between two or more objects against any standard shade, can be quantified to numerical values of colour differences in terms of total colour differences (DE values) and also in terms of DL(light and dark), Da (redness and greenness) and Db (blueness and yellowness). This has made easier to match/compare colours of textiles of a nearly matched two dyed textile samples(Standard shades given vs. Produced shade) by determining what are the colour differences between the two, leaving behind opportunity by batch correction either manually or instrumentally [1–3] to help different dyers to add or substract particular colour to get better match.

2.3.1 Batch Correction in subsequent iteration to improve match precision

The reformulation of batch correction issue for computing the incremental value of concentration of dyes by each iteration at each stage may be represented by the following matrix as follows [4]:

$$\Delta C_1 = \frac{\partial C_1}{\partial X}\Delta X + \frac{\partial C_1}{\partial Y}\Delta Y + \frac{\partial C_1}{\partial Z}\Delta Z \tag{19}$$

$$\Delta C_2 = \frac{\partial C_2}{\partial X}\Delta X + \frac{\partial C_2}{\partial Y}\Delta Y + \frac{\partial C_2}{\partial Z}\Delta Z \tag{20}$$

$$\Delta C_3 = \frac{\partial C_3}{\partial X}\Delta X + \frac{\partial C_3}{\partial Y}\Delta Y + \frac{\partial C_3}{\partial Z}\Delta Z \tag{21}$$

2.4 Colour measuring instruments

Photometry is the most common analytical technique used in measurement of colour in solution/solid in the laboratory. It is designed to measure the intensity of transmitted /reflected beam of light through the coloured solution/solid. Different types of instruments with Photometric principles are applied to the different ways of analysis of coloured materials /substrate/solution by different techniques [5] as listed below:

a. Where absorbed or transmitted light is measured:

- Colorimeter
- UV–VIS Absorbance Spectrophotometer

- Atomic absorption spectrophotometer, and
- Turbidometer

b. Where emitted light is measured: Flame photometry

c. Where reflected light is measured: UV VIS Reflectance Spectrophotometer

In analytical chemistry or in textile analytical chemistry laboratory, colorimetry based on UV VIS Spectrophotometer technique is used to determine the concentration of coloured compounds (analytes) in sample of a coloured solution or in solid coloured sample of textiles or leather or plastics or polymer film or any other chemical compounds" at visible spectrum of light (400–700 nm as VIBGYOR as described in **Figure 6**), is observed in the visible spectrum of electromagnetic radiation, emitted in the form of dominating wave lengths emitting VIBGYOR wavelengths ranging from 400 nm to 700 nm. Similarly UV Light /sunlight radiation usually is observed to have UV radiation of 180 to 380 /400 nm wave length, as UV-A, UB-B, and UV-C type.

Emitted/Reflected COLORS and Their Dominating WAVELENGTH (λ in nm)

The Ultra-violet range of wavelength lies betweern <190- 380/400nm, while visible spetra of ordinary light (what is considered as visibly seen to be colored materials) have specific predominated hues as mentioned below against particular ranges of visible wavelength (nm) for particular hue/colour :

Violet : 380 – 435nm
Indigo/ Blue :436 – 480nm and Greenish-Blue :481 – 490nm
Bluish-green : 491 – 500nm
Green : 501 – 560nm
Yellowish-green : 561 – 580nm and
Yellow : 581 – 595nm
Orange : 596 – 650nm
Red : 651 – 700 nm and beyond sometimes up to 780 nm

while , Near Infrared is > 780 nm and far infrared are beyond up to 2500 nm

Figure 6.
Major wavelength from incident white light dominating at visible range.

UV VIS Spectrophotometer technique of colour measurement are done by following two principles:

a. UV VIS Absorbance Spectrophotometry: (for determining UV–VIS wavelength scan pattern of a particular coloured compounds/dyes and to determine concentration of pure coloured compounds/dyes (analytes) in sample solution(dilute solution)

b. UV VIS Reflectance Spectrophotometry: **(for determining UV or VIS wavelength scan results of a solid materials surface (dyed/printed coloured textiles, leather /paper, cosmetics etc) of any particular coloured materials /dyes etc and to determine concentration of surface colour strength of any of the said coloured materials' surface for a solid sample(non- destructive)** [6].

This chapter mainly covers the basic principles of analysis of surface colour parameters vis a vis other appearance properties of the surface of solid textile materials and associated colour difference parameters by using Reflectance Spectrophotometer in terms of specific 1931 CIE and 1976 CIE formulae [1–3] for determination of all specific Surface colour parameters of textile materials, changes with or without different chemical processing and computer aided colour match prediction theories and practices [7, 8].

2.5 Reflectance spectrophotometer

Reflectance Spectrophotometer is an human eye simulated UV–VIS double beam spectrophotometer instrument for measuring colour of solid textile surface under standard illuminant and standard observer with or without Uv Absorption included and Excluded to measure reflectance, surface colour strength and to compare the colour differences of two sets of samples nearly to match or unmatched/matched coloured samples and also for storing and analysis of database of colour values of different textile dyes applicable to different textile substrate and its use for computer aided colour match prediction at finger tips with specific allowable limits of ΔL^*, Δa^* and Δb^* and ΔE^* values of colour differences as well as Studies on quality checking of Dyes, effect of dyebath additives, dyeing process variables and UV absorbance criteria of dyes / additives and effect of UV-absorbers/Optical brighteners on textile fabric under given treatment conditions and after care treatment/washability etc.

2.6 Diagram of Relecatance spectrophotometer and its working principles

In UV–VIS reflectance spectrophotometer based on measuring reflectance of the solid sample for measuring different colour parameters for quality control and colour matching purpose, there is tungsten light source, which acts as source of monochromatic incidence light and it falls on the opaque surface of the solid sample at a particular incident angle to reflect at the same angle for specular reflectance component and also reflects the incident light all around in diffused form at all angles for non-specular reflectance inside the integrating spehere. THere are specular component in and out arrangement with UV component in and out arrangement for specific requirement of setting ofv the instrument. Led detector situated in the inside circumference of the integrating sphere detects the total reflectance values at all diffused angles as well as at specific specular reflectance angle and the amount of reflected light intensity is measured and shown as the reflectance values

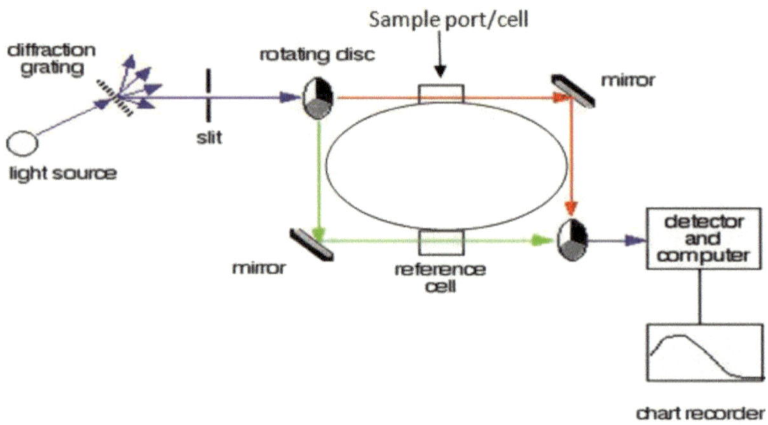

Figure 7.
Schematic diagram of working of UV–Vis spectrophotometer.

of sample at different wave length. Finally from reflectnace values obtianed for any sample, it calculates other colour parameters as per CIE −1976 formulae [1–3] and other formulae as per software inserted/installed for it for the data processing in a suitable computer system for computer aided colour measuring and also for textile match prediction system using pre-fed data base of textile dyes.

Usually a double beam UV VIS reflectance spectrophotometer, the incident light beam is first split into two parts by a half mirror as two light beam called double beam. One light beam falls /passes on the sample mounted and the other light beam falls (for reflectance mode) /passes (for transmission mode) through a control sample panel. This system of double beam eliminates the problems of interference from control sample and normalises the variations in reflected /transmitted light intensity readings uncreasing accuracy of the instrument reading, as the final reflectance/ transmittabce/absorbance values are taken as the differences between the readings of two reflected/transmitted beams of light intensity recorded. The semicircular LED Detector inside the integrating sphere measures both the two reflected/transmitted light intensity alternately and gets its processed in computerised processor to give final reading. However, in some UV–VIS spectrophotometer, a second detector is separately installed to measure the intensity of the two beams separately. Thus, the major instrumental parts of an UV–VIS Double Beam Reflectance Spectrophotometer are shown in **Figure 7** indicating the position of light source, diffraction grating, monochromator, sample cell/ integrating sphere, detector and integrator and computerised recorder. The instrument changes the light source from visible to UV light at about 350 nm by a mechanically moving mirror, as shown in **Figure 7**.

The diffused reflection shows total effect of incident light including specular reflection in integrating sphere diffraction, which may be excluded by opening a port at particular angle without detecting the specular reflections in UV–Vis reflectance spectrophotometer and different types of solid samples with varying surface and texture show variation in reflectance values, effecting surface colour strength, as shown in **Figure 8**.

Sphere Geometries of illumination and viewing in reflectance spectrophotometer [7, 8] is very important here. It is based on mesaurement of Reflectance of dyed samples of textiles. On a glossy surface there are mirror-like (specular) reflections and there are more reflections in the case of diffuse light sources. **Figure 9** shows the effect of transmission mode and total reflection mode of integrating sphere of in

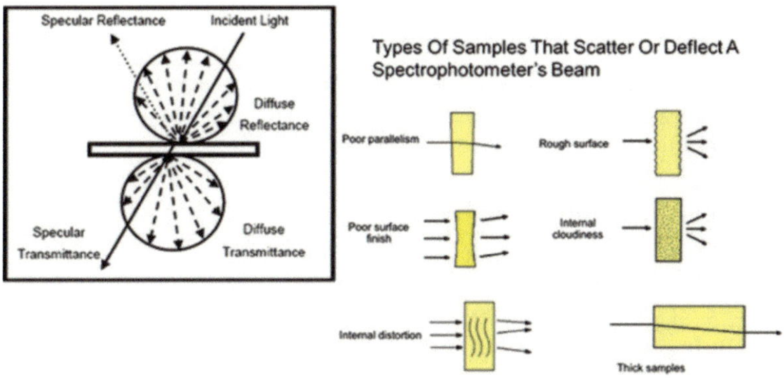

Figure 8.
Schematic diagram of optical system of UV–Vis reflectance spectrophotometer.

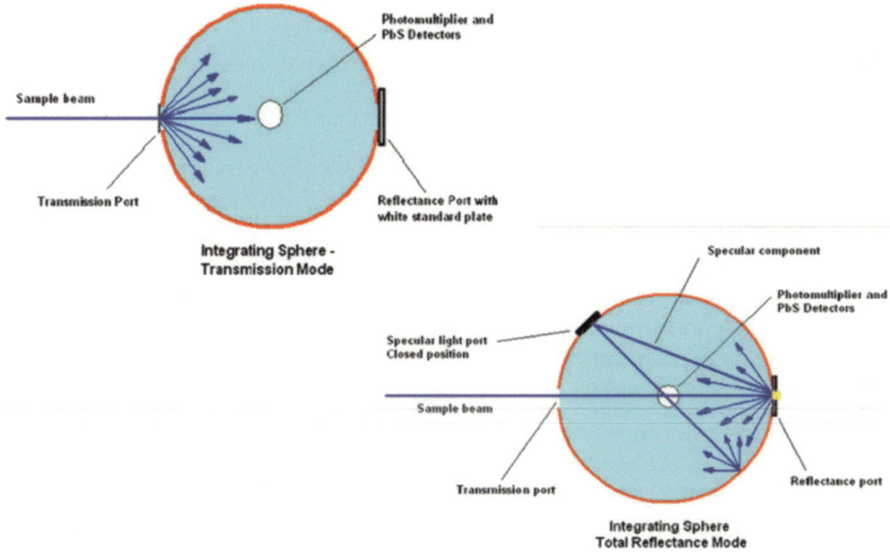

Figure 9.
Schematic diagram of working of integrating shere type optical system in UV–Vis reflectance spectrophotometer.

UV–Vis reflectance spectrophotometer, showing provision of specular component inclusion and exclusion by keeping specular port off and on (close or open).

Since the colour of the illuminant is white, specular reflections add white, with the effect of de-saturating the colour. Textiles or any non-metallic glossy surfaces look more saturated in directional than in diffuse illumination, while matte surfaces scatter the light diffusely — matte surfaces usually look less saturated than glossy surfaces.

Most of the textile surfaces are between glossy and matte and hence in reflectance spectrophotometer, diffuse illumination is provided by integrating spheres with provision of gloss traps /lid at regular reflection points to include or exclude specular/regular reflectance in instrumental set up. REflactance spectrophotometer Instruments with 45/0 and 0/45 geometry are less critical and give better results and accuracy. ASTM recommends [1–4] use the geometry that minimises surface effects (usually the one that gives lowest Y and highest excitation purity) for partly glossy samples. 45/0 geometry gives rise to polarisation problems [9].

2.6.1 Calibration and certification of the instrument 's accuracy

White calibration: Before Use always this instrument should be callibrated with standard white tiles equivalent to white surface of saturated dry layer of Magnesium sulphate, which adjusts computational parameters for any setting disturbances, so the calculated reflectance values match with calibrated value of white tiles being 100 and the accuracy of reflactance curve is the same as the absolute reflectance curve do it often.

Absolute certification: The instrument need better to be certified by NABL, which verifies that the measured colour of the standard white tile is 100 or specified values by manufacturer, within the tolerance (e.g. within 0.6 ΔE^* units) from the absolute colour of the standard white tile, in perfect agreement between instrument and laboratories, when checked for certifiocation.

Relative certification: verifies if the measured colour of the standard white callibrated tile is within the tolerance (e.g. 0.3 ΔE^* units) from the initial colour of the standard white tile in the same instrument, which is very very important for reproducibility and reliability of colour data produced.

2.6.2 Measurement of overall colour difference index

Colour Difference Index (CDI) [10] indicates the combined effects of different known individual colour difference parameters between any two samples when dyed with varying conditions of dyeing, indicating dispersion of colour value, to understand the combined effects of different dyeing variables by a single parameter. For application of same concentration of dye between two sets of dyeing under any varying conditions of dyeing like pH, taking only the magnitudes of the respective ΔE, ΔC, ΔH and MI values (irrespective of their sign and direction) may be considered to calculate CDI values using the following empirical relationship [10].

$$\text{Colour Difference Index (CDI)} = \frac{\Delta E \times \Delta H}{\Delta C \times \text{MI}} \tag{22}$$

Higher the CDI value dispersion of Colour values are more widely dispersed and that variable become critical for reproducibility for such dyeing. So, Lower CDI value below 5.0, is considered as good.

2.6.3 Use of Standrad Illuminats in reflectance spectrophotometer

The followings are the Standrad Illuminats [1–3] used in Reflectance Spectrophotometer, providing UV -Tungsten lamp for different illuminant with swift arrangement of Illuminnat -A to Illuminnat -D65.

- **Standrad Illuminant A (CIE)**: CIE standard illuminant -A is defined as equivalent light source from a tungsten filament (as radiator) when heated at 2854°K(as correlated colour temperature)

- **Standard Illuminant D65 (CIE)**: CIE standard illuminant -D65 is defined as a representation of natural daylight considering as equivalent light source from a tungsten filament (as radiator) when heated at 65040 K (as correlated colour temperature)

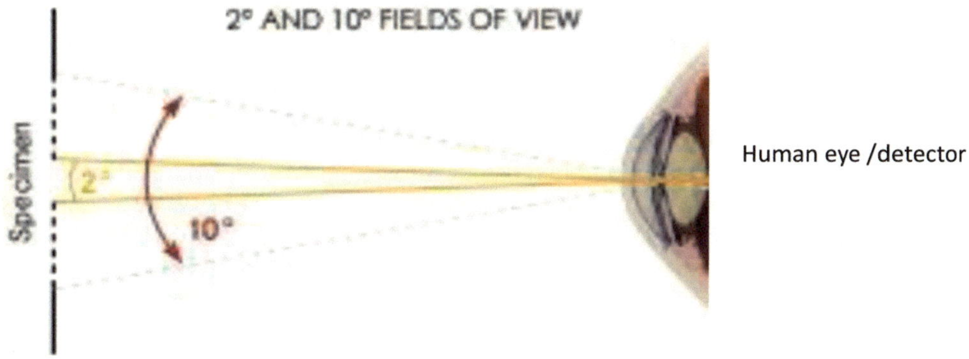

Figure 10.
2^0 standard observer (small area of view) and 10^0 standard observer (large area of view).

- **Standard illuminants B and C**: CIE Standard illuminant -B and C are defined to represent as simulated direct sunlight, as equivalent light source from a tungsten filament (as radiator) when heated at 4874°K and 6774°K respectively (as respective correlated colour temperature). But, Standard illuminants B and C are not so much used and are being dropped because they are seriously too much deficient in the UV region (important for fluorescent materials)

- **Standard Observer:** There are two angular view areas considered as standard viewing areas called 2^0 standard observer (small area of view) and 10^0 standard observer (large area of view) as shown in **Figure 10** [1–3]. As recommended by CIE, in 1964, the larger area of view of solid samples mounted for colour measurement on sample port of UV VIs reflectance spectrophotometer is most widely used for evaluating colour values and for colour matching of various types of solid samples including textiles. Ordinary Colorimeters, on the other hand, typically use a 2^0 Standard Observer (as per CIE, 1931), which has a smaller area of view and is common for general colour measurement and colour quality control and evaluation purposes for comparative purposes and also for printed textiles.

2.7 Computer aided colour match prediction system

Computer aided Colour Match prediction system (CACMPS) [4, 9] is the combination of specific hardware and software for scientific use for measuring colour of solid textile surface for given sample as standard for predicting the dyeing recipe or formulation for the exact shade reproduction in a textile fabric sample to produce. Hence, this technique is known by names e.g-computer colourant formulation, computer recipe prediction, Instrumental colour matching system or Computer aided Colour Match prediction system (CACMPS) using reflectance spectrophotometer and associated computerised sytem for storing and analysis of data with specific software to predict colour matching of textile substrate. A colour matching computer system consists of the following three basic modules,:

1. Colour measuring instrument: A Reflectance Spectrophotometer with specific geometry of colour measurement, which expresses the colour in numerical form in terms of X,Y, Z or R or K/S values with L^*, a^* and b^*, ΔL^*, Δa^* and Δb^* and ΔE^* values.

2. Computer hardware: Usually latest PC or Laptop based Computing and data analysis and storing system for data storing, analysing and processing for converting and comparing etc.

3. Specific Computer aided colour match prediction software with desirable Logic system for computer aided colour measuring /storing and analysis of colour data to convert into relevant information in terms of calculating X,Y, Z, R, K/S values and L^*, a^* and b^* values or, ΔL^*, Δa^* and Δb^* and ΔE^* values for textile surface colour measuring and colour match prediction by the said / compatible specific Software to obtain desirable output required.

So, Suitable Software is crucial in recording, analysing for colour measurement and matching and for comparing a pair of nearly matched textile dyed/coloured samples. Samples should have the same type of fabric, same surface finish and same shape as far as possible, for accurate measurements.

The Reflectance Spectrophotometer has small view and large view sample mounting holders with small hole area or large area hole respectively to place sample to scan its surface for colour measuring and recording in the diode type detector to use for storing and analysis for comparison to obtain X,Y, Z or R or K/S values with L^*, a^* and b^*, ΔL^*, Δa^* and Δb^* and ΔE^* values for further use and processing for computer aided colour match prediction from stored colour data of dyes known as database (Fabric type wise, Dye class wise or Dye Company wise etc) as per requirement.

Standard Name		=	**C51555ton cotton**				**Mode**	=		**Spectro % RFL**
3.28	3.50	3.83			4.77	5.94	8.27	9.61	11.71	
12.19	'13.08	10.72			9.42	8.67	7.68	6.53	5.36	
File	**name**	= **Dir cot**	—	**3**	**dye combination**					
ID# selected		= **2, 3,**	**4,**	**,5,**	**6, 9, 11,**					
Subst ID#		= **1 bleach cotton**								
DED65'. 1.00		**DEA. =**	**1.00,**		**Amount**	**t = 100**				
Exhaustion factors are included, Operator										
ID# Colourant		**Amount**	**Per cent**			**da***	**db***	**dL* CIE**	**dE***	**Rs**
Recipe Generated										
Formula#1										
3	**OrangeSE**	0.0408	0.0408			D −0.0	−0.0	0.0	0.0	18.51
5	**Cry SE**	0.4238	0.4238			A −0.6	0.2	0.0	0.2	187.73
9	**Turquoise Blue**	1.6111	1.6111			F −0.5	0.3	0.3	0.7	285.44
		2.0747	2.0747							491.68
Formula # 2										
4	**Scarlet**	0.0362	0.0362			DS −0.0	−0.0	0.0	0;0'	16.12
5	**Cry SE**	0.5466	0.5466			A −0.5	0.3	0.0	0.6	194.77
9	**Turquoise Blue**	1.4834	1.4834			F −0.4	0.2	0.1	0.5	224.04
		2.0662	2.0662							434.93

Table 1.
A case study of colour match prediction from the database of direct dye for cotton.

After that, the associated Computer aided colour match prediction software takes over the rest part of the work of calculations and comparison of colour data to show the measured values and calculated colour values using stored colour database of specific dyes for specific type of textile substrate. **Table 1** shows Computer aided colour match prediction system (CACMPS) generated dyeing recipe and its dye cost and estimated approxmiation of Colour difference values with least metameric ratio for cotton fabric sample to be dyed to match against given standard samples using direct dye data base stored.

The above shown formulae of colour match prediction generated against standard shade C5, has generated two possible recipes, and is difficult to decide which one we should accept and proceed for bulk dyeing. From the Point of least metamerism, formulation-1 and from the point of least cost Formula −2 are respectively found better after 4 trial run in Computer aided colour match prediction system (CACMPS) within DE limit to 1.00. Thus, computer predicted formulation −2 is least cost and formulation −1 is least metameric in nature, as shown in output result generated here.

3. Some practical consideration during measurement on CCMP system

The practical aspects of data base generating match generation using a data base, setting up proper DE or multiple colour tolerance & above all accurate spectrophotometer measurement depends on following factors.

3.1 Colour matching tolerance set-up and pass/fail system

It is also essential to develop mutually agreeable pass/fail system between buyer or seller or by company itself for their shade control to match produced samples lot with the colour values of standard shade given, which should be specified by set up specific tolerance values of these colour difference criteria for any par of near matched textiles. These may be more important while considering batch to batch variation during production of shades as per match of standard samples given. As all textile dyers and dyeing units or composite textile mills have procured this computer aided colour match prediction system, this become a regular job to check match accuracy from shift to shift or lot to lot variation always, to understand the colour differences from standard shade given. So that the colour differences from batch to batch variation at factory /dye house in company production department, the shade may be corrected by revised addition of dyes, to obtain more precision match, so that chance of rejection in export level on this ground may be eliminated and for this, specific sets of colour tolerances values are decided and pre -set.

In practice in textile industry/ dye houses, for dyed /coloured cotton textiles, if not otherwise mentioned/specified, the thumb rule for setting a symmetric colour tolerances values in terms of dL*, da*, db* and dE* for effective colour matching of cotton textiles are as follows:

Standard set of colour tolerence values: dL* = 0.7 to 1.2; da* = 0.6 to 1.0 db* = 0.6 to 1.0; dE* = 1.0 to 1.5 applicable for selected common shades and common dyes for cotton. However, these colour tolerence limits for colour matching of dyed textiles could be somewhat narrower in case of requirement of precission matching.

The main dependant factors responsible for lot to lot or batch to batch dyeing production with variation of shade are: i) weighing / solubilisation/dilution error, ii) substrate and pre treatment variation, iii) pretreatment or heat setting variation, and iv) Variation in dyeing conditions by change of dyeing process variables or variation in additive concentrations and 5) dye selected and its purity.

3.2 Calibration dyeing for preparation of data base

The calibration dyeing for preparation of dyestuff data base for dyeing specific textile fabrics with selected type and class of dyes is an essential pre requisite. After selection of substrate and class of dyes and dye manufacturer/suppliers), to run a computer aided colour measuring and matching system, preparation of dyestuff data base ton store using selective class and type of dyes, the control bleached cotton/ otherwise textile fabric sample are to be dyed with each selected dye at 5-8 different concentration levels of dyes (say, within 0.1 / 0.5 to 4%) and those samples dyed are to be subjected to measure their reflectance values at different wavelength and their individual values or their integrated sum of these REflectance / tristimulus colour values are to be stored for futurev uses to form a Dye class wise/company wise data bank or data base of all different type of dyes for specific substrate / substance following a particular standard process of dyeing in a separate file of the computerised processor or computer to use at every re-call. Samanta et al. [9, 10] mentioned the cares necessary for accuracy in colour measurement of textiles including nos. of folding etc. and orientation of samples and measure of colour difference index values etc. Randall and Stutts [11] specifies how to prepare reliable samples of calibration dyeing for creating dyestuff data base in computer aided colour matching system as a most important step. For optimum efficacy in computer aided colour matching system, the laboratory dyeing machine and process must be highly controlled in terms of dyeing process variables and all these must be standardised before callibration dyeing be carried out accurately, which is to be assured by the lab dyers /colourist to store precision colour data/ dyestuff database separately for company wise /substrate wise for separate class of dyes.

3.3 Checking of linearity / non-linearity for plots of K/S vs. dye conetrations in dyestaff data base

The accuracy of computer aided colour matching system depends on the correct dyestuff data base preparation as discussed in calibration dyeing in item 3.1 above. The accuracy of dyestuff data base can be checked by checking linearity of K/S vs. Dye Concentrations curve pattern for each individual dye applied on same substrate under standardised control dyeing conditions. Sometimes, this linear relation does not exist and then the deviation from linearity is to be eliminated, before such dyestuff data base are stored for future use of colour matching functions. The deviation from linearity of plot between K/S vs. dye concentration are due to (i) inherent variation in dye uptake rate or variation in exhaustion rate of the dye for higher percentage of dyes (ii) unknown interference of dyes with given dye bath auxialaries (iii) variation in dyeing conditions /stirring rate (iv) weighing/solubilisation/dilution error (v) impurities/agglomeration of few dyestuff itself, where these said reasons causes a variation of dyeing with increase in dye concentration [9] showing non linearity/deviation from linearity in observed. Dye uptake.

Therefore, this linearization is to be ultimately done by statistical best fit linearization process or elimination of one or two concentrations of dyes(where / from which point the said linearization is originated/deviated) for particular dye or by empirical modifications of the equation of K-M functions (K/S value). before storing dyestuff database to be used for colour match prediction of coloured textiles easily. For this type of special cases, the dye absorption co-efficient/difussion coefficient of the dye is to be determibned and to be checked at about five to eight level of dye concentrations to check the dyeing absorption/diffusion rate and then linearization can be made either eliminating few dye concentrations where dyeing rate is much varying. Repeat colouration is to be done to avoid variation in dyeing process variables to get correct data. Only after linearization of plots of K/S vs. Dye concentrations, the individual dyestuff data base may be stored in the computerised storing and saving file as ready made database for use in predicting recipe for colour matching formulations for specific substrate for specific dye class within user choiced/ standard tolerances of colour differences in terms of DE, DL, Da, Db values under standard illuminants of D65 or otherwise. If dyestuff cost are entered and uploaded and updated regularly, the dye cost of predicted colour matching formulations are also available along with predicted metamerism values.

3.4 Effect of variability in measurement of colour value of textiles

Colour Matching of textiles is very much dependant on the Pigment /dye Database created –dye class based and type of fibre/fabric based and dye company based to be pre-up-loaded in spectro. To match full strength of colour, Light and dark i.e. white and black reduction are very important.

Pigments /dyes should be thoroughly dispersed and uniformly dyed, which is Very difficult with powder pigments, but much easier with Master batch mass pigmentation to produce coloured textiles.

There are so many variations in measuring surface colour of textiles. A measurement is never perfect. The effect of variability of colour measurement is reduced by using multiple measurements and taking avrages at 10 points atlesat. How many measurements should I make for averaging is a good question and Rule of thumb is 10 times atleast for each variability parameter of dyeing for standardising dyeing process variables. For any variable instrumental factors also, measure each spot of colour value for 10 times to take average of it. But for sample uniformity for data base storing data, one should repeat colour measurements at several locations — more than 10 to 100, depending CV % of K/S values or reflectance values of coloured textile surface. One can follow ASTM standard E 1345-90 to determine how many measurements are necessary in each case.

Some more Practical Aspects of variability in colour mesurement of textile surface [9] are:

a. Level / Un level dyeing (Usually Less than 5% CV of K/S Value is taken as level dyeing for textiles).

b. Back ground opaqueness of the sample (No. of Folds are to be kept Constant say usually 4 fold).

c. Variation in warp wise or weft wise sample's vertical/horizantal orientation may differ K/S value)

d. Variation in texture or surface roughness may vary K/S values for change in scattering (Any chemical/ physical intervention/Treatment before dyeing may change surface texture)

e. Variation of colour and texture in two sides of the fabric sample (K/S -surface colour strength in one face of fabric may sometimes differ from other face due to the said effect).

f. Any Fabric or Dyeing Defects/Stains in the fabric sample (Any defect of the fabric may cause colour variation).

g. Dull shade / Fluorescent colour & bright shade etc. some times behave differently.

h. Blended fabric may pose problem with change of Blend % of each component of fibres.

The reasons of variation of colours produced in textiles during data base preparations - are

- Improper weighing and mixing of colourants.
- Improper Cleanliness of dyeing machinery parts, like steam pipes and dye bath
- In-compatable colour mixing and variation of dyeing time and other process variables
- Interference from regrind/ slubilization of dyes having some chnaces of contamination.
- Shedding of fibres / Degradation of fibres or chemicals used during processing.
- Machine stoppages and inadequate steam purging in dye bath having in adequate temperature
- Improper selection of Colourant/ Master-batch.
- Interference with processing additives - chrome pigments containing lead will discolour if Tin stabilisers are present.

Moreover, different other cares are necessary, without which in-accurate measurement occur for measuring of colour values of textiles –e. g.

- Accent on cleanliness: Poor housekeeping results colour contamination and stain,
- Selecting correct Colourant and clear understanding of colour type and properties to specific textile fibre / polymer involved. Master-batch of colour supplier plays crucial role in this case.
- Use of pre-coloured standard materials and best checking by replicating same colour on a same or different textiles. Cost of instrument is usually more
- Inventory and logistics issues are there also for variation of colour and its measurement.

Use of Pigment colouring for small quantities where Master-batch development is easier.

3.5 Areas of application of computer colour measuring and matching system

1. Measurement of tristimulus values, reflectant at maximum absorbance wave length or K/S measurement of transmitants.

2. Calculation of colour difference by CIELAB equation of total colour difference i.e. ΔE* = [(ΔL*) 2 + (Δa*) 2 + (Δb*) 2]½ with plot of CIE colour difference space diagram. The shade sorting i.e. pass/fail mechanism of quality control of dyed shades from batch to batch or lot to lot represents colour difference values in terms of Darkness and lightness (ΔL*) Redness & Greenness (Δa and Yellowness and Blueness (Δb*).

3. Finger Tips solution of predicting newer computer aided d colour matching recipe /formulation with lowset cost & lowest metameric match recipe using pre- set stored speicific dyestuff data base.

4. Batch correction or auto correction of shades by manual or computerised corrections.

5. Extension of match Prediction by batch corrections and utilisation of dyeing waste liquor.

6. Purity/ Quality test of incoming newer batch of dyes with the help of this computer aided colour measuring system.

7. Determination of whiteness, yellowness and brightness indices of bleached textile substrate using different standard scales like CIE scale, Hunter Lab scale, ASTM-E-313 scale, Stansby scale etc.

8. Prediction of efficiency of OBA (optical brightening agents) utilising this system by change of with or without UV light setting.

9. More accurate and Quantitative understanding and grading of colour fastness grade for colour fading behaviour to wsah, crocking/rubbing. Exposure to UV light etc. replacing conventional grey scale rating.

4. Surface appearence measurement in terms of whiteness, yellowness and brightness indices

Whiteness is assesment of freedom from any colour and contamination/stain / soil and as such it is taken as an indicator of quality for a bleached textile fabric prepared for dyeing. Objective measurement and meaningful numerical expression whiteness index as per CIE and Hunter lab scale are widely used. It represents whiteness index (WI) in terms of colorimetric values for the specimen and the chromaticity coordinates of the illuminant:

$$\mathrm{WI(CIE\ scale)} = \mathrm{Y} + 800(\mathrm{x_n} - \mathrm{x}) = 1700(\mathrm{y_n} - \mathrm{y}) \tag{23}$$

$$\mathrm{WI(Hunter\ Lab\text{-}Scale)} = \mathrm{L}^* - 3\mathrm{b}^* = 10\sqrt{\mathrm{Y}} - [21(\mathrm{Y} - \mathrm{Z\%})]/\sqrt{\mathrm{Y}} \tag{24}$$

where x, y and Y are the colorimetric values for the sample under illuminant D_{65}, and x_n and y_n are the chromaticity coordinates of the light source/illuminant used. A value for WI of 100 represents a perfect reflecting white diffuser, equivalent to surface of saturated paste of Magnesium sulphate.

X, Y and Z are the CIE tri-stimulus values of the sample and L^* is the lightness/ darkness indicator, b^* is the blueness/yellowness indicator.

Similrly, yellowness indices [12] as per ASTM-E313/1973 can be expressed as follows:

$$\text{Yellowness Index (YI)(ASTM-E313/1973)} = 100\ \left[1\text{–}(B/G)\right] = 100\left[1 - \frac{0.8477}{Y}\right] \quad (25)$$

Where, X, Y and Z are the CIE tri-stimulus values of the sample, L^* is the lightness/darkness indicator, b^* is the blueness/yellowness indicator and $B = Z/1.181 = 0.847\ Z$, $G = Y = L^2/100$.

Brightness Index (BI) as per ISO-2469/2470-1977method [13] can be calculated by following formula:

$$\text{Brightness index} = \frac{\text{Reflectance Value of the Sample at 457 nm}}{\text{Reflectance value of Standard white diffuser (white tiles)at 457 nm}} \times 100 \quad (26)$$

Treatment with fluorescent brightening agents can lead to reflectance values of up to 150. Although the pattern appears to become whiter, the change in appearance is due to a change in chroma towards blue, and this fact is expressed in quantitative form as the ‘tint factor’. Allied to the appearance of the uncoloured fabric is the yellowness, which suggests yellowing by chemical treatment or by heat scorching or degradation by light or by gases. Thus along with colour parameter the said other surface appearance properties are also very very important too in defining the quality of the surface appearance of any textiles.

5. Importance of colour measurements and matching in garment industry

Colour is one of the important element of a design. Colour with aesthetics / texture of anty textile fabrics / garments are as important as its physical and functional property criteria. Matching of colours, especially in specific textiles made from specific or different fibres and their blends is very very crucial in many applications. The task of communicating and measuring and matching of colour becomes more difficult when colours need to be exactly matched with a given standard supplied for different textiles. More and more precision colour matching is required in specialised textile products like defence dress materials, school uniform etc. and also for matching suits for consumer textiles and livestyle products for matched furnishings, bed linen and auto Interiors etc.

This Computer aided reflactance spectrophotometer is an impoartant tools/ intrument for quality up grdation of textiles and garments by maesuring surface colour strength and colour dierences values as per CIE equations/ formulae. Some Other Application of computer aided colour measuring cum matching System used in textiles or apparel industry’s Dye House:

1. For quality control of dyed textiles including pass/fail decision of batch to batch checking.

2. For Evaluation of Quality of dyes supplied.

3. Effect of dyeing additives by measuring colour yield.

4. Efficiency of optical brighteners by UV analysis.

5. Soil removal efficiency of surfactants by measuring Reflectance value.

6. Measurement of whiteness / yellowness / brightness index etc. of undyed and bleached samples besides dyed samples.

6. Concluding remarks

Computer Aided Colour Measuring and Match Prediction System (CACMPS), now a days, become an essential tools for each textile dye houses to match colours or shade as per panton shade nos. or as per given samples, to reduce export rejection for colours. Moreover, to judge colour fastness grading more accurately from measurement of colour difference values after corresponding fading by wash or light or rubbing etc., than subjective/comparative judging by grey scale rating purpose is more scientific and advantegios. Quality control activity and batch to batch pass / fail checking of shades developed from shift to shift needs to be implemented in all dye houses for quality assurance on colour matching which is an integral demand of today's apparel and fashion industry. Hence learning of colorimetric principles of UV VIS reflectance spectrophotometer and proper utilisation of this instrument carefully for deriving all round benefits out of it, for surface colour measuring and matching of textiles for customer satisfaction is also helps in brand building by quality assurance on colour matching of textiles.

Acknowledgements

The author is thankful to the Principal RBGC, Kolkata for her encouragement and support.

Author details

Pubalina Samanta
Department of Fashion and Apparel Design, Rani Birla Girls' College, Kolkata, West Bengal, India

*Address all correspondence to: pubalina@gmail.com

References

[1] Shah HS, Gandhi RS. Instrumental Colour Measurements & ComputerAided Colour Matching for Textiles. Amhedabad, India: Mahajan Book Distributers; 1990. pp. 76–116

[2] Gangakhedkar NS. Part II : Colour measurement and its applications. In: Gularajani ML, editor. Colour Measurement: Principles, Advances and Industrial Applications. Manchester, UK: Woodhead Publishing, India and Textile Institute, Series in Textiles; 2010. pp. 219–252

[3] McLaren K. The Colour Science of Dyes and Pigments. 2nd ed. Bristol: Adam Hilger; 1986. pp. 50–87

[4] Pubalina S. Fundementals and applications of of Computer aided Colour Match Prediction (CCMP) System. Latest Trends in Textile and Fashion Design Journal. 2018;**2**(5):1-10

[5] Ingamells W. Colour for Textiles-A user's handbook. Bradford, UK: Society of Dyers and Colourists; 1993. pp. 138–157

[6] Samanta AK. On-line colour control (for textile dyeing). Indian Textile Journal. 1992;**105**:70

[7] Samanta AK. Review on Colour Control and Matching. Indian Textile Journal. 1995;**108**:50

[8] Samanta AK. A review and case study of computer application in jute or textile mills' dye house and computerized colour matching. Textile Trend. 1993;**36**:37

[9] Samanta AK, Das D. Studies on Quantitative colour measurement of direct dyed Jute Fabric in relation to computerized colour matching. Journal of Institution of Engineers-Textile Engineering. 1992;**73**:53

[10] Samanta AK, Agarwal P, Singhee D, Dutta S. Application of single and mixtures of red sandalwood and other natural dyes for dyeing of jute fabric: Studies on colour parameters / colour fastness and compatibility. Journal of the Textile Institute. 2009;**100**(7): 565-587

[11] Randall D, Stutts T. Optimizing calibration dyeings for computer color matching. American Dyestuff Reporter. 1988;**28**:44-46

[12] ASTM-E-313-67/E-313-73 (Re-approved-1979). Determination of yellowness. In: Annual Book of ASTM Standards E 12.02. Philadelphia, Pennsylvania, PA: American Society for Testing Materials; 1974. p. 1

[13] ISO-2470-1977 (E) and ISO-2469. Determination of Brightness. Switzerland: International Organization for Standardization (ISO); 1977. pp. 1–3

Chapter 6

Colorimetric Measurement and Functional Analysis of Selective Natural Colorants Applicable for Food and Textile Products

Deepali Singhee and Adrija Sarkar

Abstract

Colouration of textiles as well as food products with natural colorants is an interesting subject with respect to the growing eco-concern among the consumers. Several colorants are available in nature for textile colouration and are renewable, biodegradable, and eco-friendly. Being safe for human consumption, they can serve the dual purpose of also coloring food. Several such natural dyes are available. This review chapter deals with the chemistry, extraction, application, and colorimetric analysis of colorants derived from turmeric (root), annatto (seeds), and cochineal (insect) for use on both textiles and food products.

Keywords: eco-friendly colors, extraction of natural colors, food colors and additives, natural colorants for textiles, natural colorants for food, colorimetric estimation

1. Introduction

From the application in textiles, uses of natural dyes also extend to colouration of food and in other areas like medicines, cosmetics, and procession of leather products. Several sources of natural colorants used in the past have been re-identified today. Many are common and play a dual role in colouration of textiles as well as food products and drinks. Some dye-yielding plants contain compounds like curcumin, crocin, bixin, carthamin, punicalagin, nimbin that are known to have therapeutic properties and are used in various traditional medicinal therapies. Their inherent functional properties like antimicrobial, antifungal, deodorizing, UV protection, moth/insect-repellent, and others allow them to enhance the value of the dyed textiles, or the colored food products. This chapter deals with some selected natural colorants widely used in the textiles and food sectors and documents their chemistry, extraction process, application, usage and properties, separately, in relation to textiles and food. Few case studies on colourimetric measurements and analysis of functional properties of natural dyes on textiles and food are also discussed.

2. Natural colorants for textile related application

Natural dyes for textiles are dyes or colorants derived from plants, invertebrates, or minerals. From the plant source, colors are extracted from seeds, roots, stems, barks, leaves, flowers, berries, and fruits. In addition to the natural vegetable coloring matter, animal/insect coloring matter like tyrian purple, cochineal, lac and kermes, and mineral coloring matter derived from ocher, limestone, manganese, cinnabar, azurite, and malachite are also used to produce natural effects on the fabrics. With the advent of synthetic dyes, natural dyes faded into oblivion. But now with several advantages like fast and durable colors coupled with replaceable, biodegradable, and fairly non-polluting nature over the synthetic ones, natural dyes are making a comeback.

Different natural dyes yield different colors–yellow (kamala seed pods, myrobolan fruit); mustard yellow (latex from the gamboge tree); yellow to orange (pomegranate rind, turmeric, and lichens); peach to brown (chestnut hulls); orange (gold lichen, carrot and onion skin); pink (berries, rose and beets); crimson to maroon (teak leaves and cochineal); orange, pink and red (madder root); red to brown (bamboo and hibiscus flower); brown (catechu bark and coffee beans); red to purple (red sumac berries, basil leaves, hibiscus flower, logwood, lac); purple (red cabbage and murex snails), blue (indigo leaves), green (sorrel roots, spinach, and peppermint leaves); yellow, gray to black (black berries, iris root, and walnut hulls) and sepia brown (octopus/cuttlefish).

Different compounds are present in natural dye sources that impart a variety of colors on textiles; indigotin (blue and purple), anthraquinones (shades of red), carthamin from safflower (red and yellow shades), naphthoquinone (orange, red, or reddish-brown shades), flavonoid dyes (yellow to greenish-yellow and brown colors), carotenoid (orange), tannins (different colors with different mordants) and curcumin (yellow shades).

3. Natural colorants for food related application

Color is the prime sensory attribute of foods and is often used by consumers as an indicator of food quality in terms of flavor, safety, and nutritional value. Food colors are dyes, pigments, or other substances that impart color when added to a food product or a drink. Such additives make the food more attractive, appealing, and appetizing; provide color to colorless foods or enhance their natural color; offset color that is lost on exposure to air, moisture, high temperature, light, and unfavorable storage conditions; and allows the consumers to identify products on sight. Thus, one of the main applications of food colorants is the modification or preservation of its visible appearance.

Food colors can be obtained naturally as extracts from natural sources, or they can be synthesized. Natural food colors are usually extracted from seeds, fruits, vegetables, leaves, insects, algae, etc., and are used both in domestic cooking and commercial food production and are available in many forms such as liquids, powders, gels, and pastes.

Among the natural food colorants, Asian spices like turmeric and saffron are used in everyday cooking; they lend an appeasing color to the food. Saffron, as a spice finds its use in biryanis and as colorants in dairy products. Caramel is mostly used to enhance flavor in deserts. Hibiscus is a commonly used bakery product and tea-based beverage to enhance the brown tint. Marigold does not have extensive use but the petals are sometimes used to enhance colors in salads. Beet juice has several applications in many beverages, dairy products, yoghurt ice cream, sauces, jams, jellies, and candies.

Different sources of natural colorants yield different colors; dark yellow is obtained from turmeric; yellow-orange from saffron; orange from carrots, red pepper/paprika, and sweet potato; pink from strawberries and raspberries; red from carrot, beets, and tomato; deep red from beetroot and red sandalwood; green from matcha and spinach; blue from red cabbage mixed with baking soda; purple from blueberries and purple sweet potato; brown from coffee, tea, and cocoa; and black from activated charcoal and squid ink.

A variety of compounds present in natural dye sources are responsible for different colors. Anthocyanins (flavonoids) found in fruits and vegetables are responsible for blue, purple, red, and orange colors. Carotenoids in fruits and vegetables are known for imparting red, orange, and yellow colors. Betalains present in most caryophyllales plants give a pink to red color. Curcumin is responsible for the yellow color of turmeric. Safflower gives an attractive yellow color. Chlorophylls from alfalfa (*Medicago sativa*) are responsible for the characteristic green color. Carminic acid in carmine from cochineal is responsible for dark purplish-brown or bright red or dark red color.

4. Colorimetric measurement of natural colourants

The appearance of a textile or food material is ascertained through its surface color and is the first sensation perceived by the consumer to judge its acceptability. The color of an opaque object is described by the reflectance of light as a function of its wavelength. The human eye is versatile and can detect light and light modification by the colorant and this is interpreted by the brain as color. For any color to be perceived by a human eye, a source of light, an object, and an observer is required.

Color measurement of products can be carried out in two ways; by visual evaluation or through instrumental analysis. The chromatic attributes and different geometric factors like texture, shape, etc. of foodstuffs can be assessed qualitatively by the human eye. In this process, the observer assesses the color of the sample under standard conditions of illumination, and after comparison with defined color standards; the assessment is defined in terms of some scores generally on a 9-point scale. One of the most popular scales is the 9-point Hedonic scale in which the products can be marked from 1 to 9 depending on the appearance and acceptability rate of the food product. A lower score indicates low and least acceptable color intensity; while a high score denotes high color intensity or acceptable appearance. Such visual assessment is subjective, relative, and is dependent on the observer and environmental conditions. On the other hand, the presence of color pigments can be also be quantitatively assessed using different types of equipment. But each instrument measures only one attribute at a time and so several instruments may be needed to measure various aspects of visual perception. Basically, there are three types of instruments that measure color or its attributes, colourimeter, spectrophotometer, and spectroradiometer.

Liquid chromatography is a method for separating, identifying, and quantifying the constituents of a mixture. The interaction of the sample with the mobile and stationary phases causes this separation. Because there are so many distinct stationary/mobile phase combinations that can be used to separate a mixture, chromatography is divided into various categories based on the physical states of those phases, liquid, and gas. Liquid–solid column chromatography is the most common chromatography technique that uses a liquid phase (mobile) that filters down through the solid stationary phase, bringing the separated components with it. To separate the components that make up a sample, high-pressure liquid chromatography (HPLC) uses pumps to push a pressurized liquid solvent containing the sample mixture through a chromatography column loaded with solid absorbent materials. Each component in the sample interacts with the adsorbent material in a slightly different way, resulting in varying flow rates and separation of the components as they flow out of the column. The type of chromatography column employed determines how different chemicals are separated. Several different types of columns (size exclusion, ion exchange, normal phase, reverse phase) are used. Once the molecules make it through the column, they will be detected by a detector, which is typically a UV detector, but other detectors such as refractive index detectors, laser light scattering detectors, fluorescence detectors, and thermal conductivity detectors are also used. High-performance liquid chromatography (HPLC) is considered the 'gold standard' for measuring pigment concentrations in plant samples. A major drawback of this process is its high cost both in terms of time required for assessment, and the high cost of the testing equipment itself. Liquid chromatography can be combined with mass spectrometers (LC–MS) to analyze organic and inorganic compounds of biological origin. While liquid chromatography may separate mixtures with several components, mass spectrometry can identify the individual components' structural identity with high molecular specificity and detection sensitivity.

Colorimetric or spectrophotometric analysis is another technique to evaluate color in textiles or food. Because the amount and color of light absorbed or transmitted through a solution is dependent on the concentration of pigment particles present in it, such measurements rely on detecting the concentration of material (color/pigment) in a solution. Such color evaluation measures the change in the intensity of electromagnetic radiation in the visible wavelength area of the light spectrum after it is transmitted or reflected by the object or solution through which it passes. A colorimeter or spectrophotometer thus assesses the color in various sample solutions (dyes in textiles, or colorants in food) by absorbing a particular wavelength of light and denotes the assessment in the form of some values using the Beer–Lambert law. Under Beer's law of photometry, the amount of light absorbed is proportional to the solute concentration present in the solution. According to Lambert's law, the amount of light absorbed is proportional to the length as well as thickness of the solution taken for analysis or in other words, when light passes through a medium, its absorption is proportional to the intermediate convergence. Beer's law and Lambert's law are usually taken in combination as Beer–Lambert law which indicates the relationship of absorbance with both the path length of light inside the sample and the concentration of the sample.

Thus, the principle of operation of a colorimeter is outlined as follows—in a colorimeter a beam of light of a given wavelength is directed toward a liquid sample (of the dyes in textiles, or colorants in food). While passing through a solution in the colorimeter, the beam of light travels through a series of lenses, and the photocell is able to detect the amount of light passing. The current produced by the photocell depends on the quantity of light striking on it; higher the concentration of the colorant/pigment in the solution, the higher is the absorption of light and consequently less transmission. Thus, less light passing through the solution would indicate the creation of less current by the photocell [1]. The colorimeter can qualitatively detect the presence of color pigment in a sample when the wavelength peak detected in the experimental sample matches with the peak (λ_{max}) of the standard pigment.

The colorimeter can also measure the amount of pigment present in the sample. In this case, calibration curves can be made using the different concentrations of the standard solution of the pigment. With the help of a calibration curve, the amount of pigment present in the sample can be estimated. In case standard solutions are not present, then various equations can be formulated using extinction coefficients, molecular weight, etc. to ascertain the amount of dye pigment in the sample.

When items are viewed under different sources of light and illuminations, their colors are frequently diverse. The discrepancy is due to differences in the spectral power distribution of the illuminations as well as changes in the lighting. An illuminant is a specific spectral power distribution incident on the object viewed by the observer, whereas a source is a physical emitter of radiant energy, such as a lamp or the sun and sky. As a result, a single source of light can provide several illuminants. Illuminants can also have a variety of spectrum power distributions. Numerical specification of color was earlier visualized by chromaticity diagram and the three chromaticity coordinates (x, y, and z) were calculated by the use of the three tristimulus values that represent the amount of standard lights (red, green, and blue) required to reproduce a color.

Over time, a slew of alternative color appearance models have arisen, as well as a numerous new color measurement related terms. To represent the color of an item, several color coordinate systems can be employed, including RGB (red, green, and blue), Hunter Lab, Commission Internationale de l'Eclairage's (CIE) L*a*b*, CIE XYZ, CIE L*u*v*, CIE Yxy, and CIE LCH. Almost of modern color measurement is based on experimental observations in accordance with the CIE (International Commission on Illumination) color specification system. The human eye has three color receptors: red, green, and blue, according to CIE principles, and all colors are combinations of these.

Color evaluation methods such as the Hunter Lab L*,a*,b* and the modified CIE system known as CIELAB are widely used in the food and textile industries. They were created as a result of investigations that correlated tristimulus values with visual perceptions of color in order to convert the X, Y, Z system (tristimulus values) to a visually uniform color-system. Each color can be considered equivalent to a member of the greyscale lying between black and white, according to L*, which is an approximate measurement of brightness. As a result, the L value for each scale reflects the level of lightness or darkness, whereas the a and b values indicate redness or greenness, respectively. Hunter L, a, b is a color scale based on the opponent-color theory which states that color receptors in the human eye see color as pairs of opposites: light–dark, red-green, and yellow-blue. To fully define the color of an object, all three values are required. The scale consists of two color coordinates, a* and b*, as well as a psychometric index of lightness i.e. L*. The parameter a* is positive for reddish colors and negative for greenish colors, whereas the parameter b* is positive for yellowish colors and negative for bluish colors. L* is an approximate measurement of luminosity according to which each color can be considered as equivalent to a member of the greyscale lying between black and white. Thus, the L value for each scale, therefore, indicates the level of lightness or darkness; the values indicate redness or greenness, and the b values yellowness or blueness. The CIE 1976 L*a*b* color or modified CIE system called CIELAB was recommended by the CIE in 1976 to improve on the 1966 version of the Hunter L, a, b. The CIELAB color scale, like the Hunter, expresses color as three values: L* for perceived brightness, a* and b* for the four distinct hues of human vision: red, green, blue, and yellow. Under the two color scales, however, three values of L, a, and b are determined differently; the formulas for Hunter L, a, and b are square roots using CIE XYZ, whereas CIELAB uses cube roots of XYZ. The CIELAB color scales were designed to be a perceptually uniform space in which a given numerical change correlates to a corresponding perceived change in color, and so provides a better approximation to the visual judgment of color difference for very dark hues. Despite the fact that the LAB space is not genuinely perceptually uniform, it is valuable in the industry for detecting minute color changes. Because the CIE L*a*b* scale, which was released in 1976, has gained popularity, the Hunter color scale is no longer as widely used as it once was. Although CIE measured the single color space, it was not truly uniform visually throughout the color space and could not define color-difference in a singular term i.e. two colors cannot be red and green at the same time or yellow and blue at the same time. It meant that equal color difference magnitude appear of different visual magnitudes in different regions of the color space. For this reason, the CMC equation (Color Measurement Committee) or color difference (ΔE^* or DE^*) formula which takes the non-uniformity of the color space into account is used to assess the difference between two colors and is more preferred in textiles color assessment today. The CMC equation corrects the CIELAB color scale's most significant flaw, which is chroma location dependency.

The total color difference, ΔE, may also be calculated. A comparison of two colors is used to determine this color difference (ΔE^* or DE^*). One is designated as the standard (or target), and the other as the sample. ΔE is a single value that takes into account the differences between the L, a, and b of the sample and standard. The delta values (ΔL, Δa, and Δb) show how far a standard and sample differ in terms of L, a, and b. Different color difference formulae are used to calculate the numerical color difference between two colors.

- ΔL^* (L^* sample - L^* standard) = difference in lightness & darkness (+ve = lighter, −ve = darker)
- Δa^* (a^* sample - a^* standard) = difference in red & green (+ve = redder, −ve = greener)
- Δb^* (b^* sample - b^* standard) = difference in yellow & blue (+ve = yellower, −ve = bluer)

Deltas for L^* (ΔL^*), a^* (Δa^*) and b^* (Δb^*) may be positive (+) or negative (−). Whether the sample is redder or greener than the standard is indicated by the sign of the delta value. For example, a sample will be redder than the standard if Δa is positive. The total difference, Delta E (ΔE^*) is always positive. For the delta values, tolerances can be established. Out-of-tolerance delta values indicate that the discrepancy between the standard and the sample is too great. If ΔE is out of tolerance, it is difficult to know the parameter that is out of tolerance. It can also be deceiving in situations when L, a, or b are out of tolerance but E is still within it.

Color values of textiles are also assessed in terms of K/S (Kubelka-Munk) values where higher values represent darker and more saturated colors. K/S values are usually calculated at the wavelength of maximum absorption of the color (λ_{max}); however, a calculation over the visible region may also be employed. The Kubelka-Munk equation is as follows:

$$K/S = \frac{(1 - R_{\lambda max})^2}{2R\lambda_{\lambda max}} \quad (1)$$

Where K: is the constant related to light absorption of the dyed fabric; S: is the constant related to light scattering of the dyed fabric; R: is the reflectance of the colored fabric that is expressed in fractional form.

The objective measurement of color is thus dependent on the quantification of the light source (E), the object's reflectance (percent R), and the observer's color response functions r-g-b. In food products, color quality is either measured on a spectrophotometer and expressed in terms of the chromatic attributes (L^*, a^*, b^*) as proposed by CIE, or in terms of tint values measured using a tinctometer and interpreted as color ratio between yellow and red pigments (R and Y values). Colors on textiles can be characterized by hue (dominant shade); the amount of color present or saturation; and by the degree of lightness or darkness of the particular color. Thus in textiles color values are generally expressed in terms of the color strength (K/S values), color difference (ΔE), chromatic attributes (L^*, a^*, b^*), as proposed by CIE and Metamerism Index (MI). Based on the respective magnitudes of ΔE, ΔC, ΔH, MI, a newer empirical index CDI (Color difference index) of assessing color for a binary mixture of dyes has also been postulated [2].

5. Some selected colourants commonly used for colouration of textiles and foodstuffs

5.1 Turmeric

Turmeric is derived from the tuberous rhizome of the Zingiberaceae family. *Curcuma longa*, the yellowish-brown rhizome from which the turmeric is derived develops beneath the ground and is cylindrical, tuberous, highly branched with a rough and segmented skin with a dull orange interior. The leaves are pointed and the flowers are funnel-shaped and yellow in color. *C. longa* is a perennial herbaceous plant that grows wild in tropical Asia. India is the largest producer, consumer, and exporter of turmeric in the world contributing 78% followed by China, Myanmar, Nigeria, and Bangladesh together contributing to 6% of the global production. Dried the turmeric rhizome gives yellow powder with a bitter, slightly acrid, but sweet flavor. *C. longa* is a medicinal plant that is used extensively in textile and food colouration. It is popularly used as a spice in South Asian and Middle Eastern cuisines as it lends a distinctive yellow color and flavor. *C. longa* also possesses antioxidant, anti-inflammatory, choleretic, antimicrobial, and carminative properties and has been used in traditional Indian ayurvedic medicine. The dye has been used to color fabrics in brilliant yellow colors. It can be used in combination with other plants like *Butea monosperma* flowers [3] or *Nyctanthes arbor-tristis* flowers [3] to give a range of yellow shades. It's typically used as a foundation color for indigo overdyeing to achieve a fast green.

Genus: *Curcuma* | **Species**: *longa* | **Family**: Zingiberaceae.

Common name: Turmeric | **Local name**: *Haldi.*

Part of the plant used for coloring: Roots/rhizomes and leaves.

5.1.1 Coloring pigment/component

Turmeric has a volatile oil that contains turmerone, as well as other coloring compounds called curcuminoids mainly concentrated in the rhizome. Curcuminoids (1,7-bis 4-hydroxy-3-methoxyphenyl-1,6-heptadiene-3,5-dione) are natural antioxidants and curcumin is the principal curcuminoid present in turmeric. The other two curcuminoids are desmethoxycurcumin and bis-desmethoxycurcumin. Curcumin is a polyphenol and the principal coloring component of this yellow dye which has been also been classified as CI Natural Yellow 3 and considered a direct type of dye. Curcumin can be found in two different tautomeric forms: keto and enol. In the solid-state and in solution, the enol form is more energetically stable [4]. The chemical structure of curcumin is different under different pH and hence it can be used as an indicator. It remains yellow in an acidic medium, while when added to an alkaline medium above pH 8, the shift of the hydrogen atom causes the compound to change color giving a red hue. It is not soluble in water (acidic and neutral pH) at room temperature but is soluble in oil and alcohol. Curcumin also has fluorescence qualities, which extends the active life of these molecules and increases the chances of contact with oxygen in the air, making them more susceptible to photochemical

Curcumin content in different types of turmeric	L	a*	b*
3.5	32.6	39.1	31.5
3.8	36.6	28.4	36.1
4.3	46.3	22.1	42.0
5.1	54.7	17.5	46.1

Table 1.
Variation in color values with respect to changing curcumin content in turmeric taken from difference sources.

oxidation. [5]. A relationship exists between the curcumin content and the L*a*b* values [6] and high curcumin content is associated with high L* (lighter) and b* (yellower) values, but with lower a* (less red) value. Where a* and b* values are high, the resultant shades are red and yellow respectively, while when both a* and b*values are similar, the resultant shade is orange (**Table 1**).

5.1.2 Application of turmeric in textile coloration

Very few studies have been reported on dyeing of textiles with turmeric. Cotton was dyed with purified ethalonic extract of turmeric by the exhaust technique [7]. Enhancement of dye uptake and wash fastness of cotton was achieved through modification with enzymes and chitosan [8], irradiation with gamma rays [9], and microwaves [10] before dyeing. Silk was dyed with *Curcuma Longa L* rhizome in brilliant shades [11] and improved dye uptake and fastness were obtained on silk pre-irradiated with methanolic extract of *C. longa L* rhizome [12]. Nylon dyed with turmeric gave fast colors [13].

5.1.2.1 Color produced

Turmeric yields a warm gold color on undyed natural cotton fabrics, silk, and wool. It gives a wide range of yellows without mordants. With mordants (metal salts), it gives colors like golden yellow (tin), mustard yellow (copper and chromium), and olive green (iron). Its wavelength of maximum absorption (λ_{max}) is 420 nm [14] or 450 nm [15] indicating that the dye can absorb color in the blue end of the spectrum. The wavelength of maximum absorption for turmeric is.

5.1.2.2 Extraction

Maximum yield (highest absorbance) of color from turmeric was obtained at pH - 6 at 100°C [16] indicating that the dye can be extracted under very mild acidic or neutral conditions. Also, maximum extraction occurs at high (boiling) temperatures [5]. The solvent extraction process gave maximum yield followed by aqueous extraction, but the purest form was obtained by spray drying [14].

5.1.2.3 Dyeing conditions

Color strength (K/S) value of the dyed fabric was maximum in pH 7 [7]. Good color strength was observed by dyeing fabric irradiated at 65°C for 40 min in dyeing bath having pH 6 [10]. Glauber's salt tends to neutralize or reduce the negative electric charge (zeta potential) of cotton fabric, thus facilitating the approach of the dye anions to the fabric within the range of formation of hydrogen and other bonds between the dye molecules and fabric and thus the color strength of cotton dyed with turmeric extract increases with increase in salt concentrations [5].

5.1.2.4 Fastness

In general, turmeric is a fugitive dye and bleeds easily. Turmeric exhibits poor washing fastness due to the phenolic groups present in curcumin which reacts with soda ash (in washing liquor) forming curcumin salt that is soluble in water and hence can be easily washed out from the dyed fabric. The poor light fastness of turmeric is attributed to the inherent susceptibility of its chromophore to photochemical oxidation. However, both the wash and light fastness of textiles dyed with turmeric can be improved through mordanting. The improvement in light fastness can be attributed to the reduced susceptibility of the turmeric dye chromophore to photochemical oxidation in the presence of mordant. Though dyeing with turmeric exhibits good fastness to rubbing, a decrease is noted both in the dry and wet rubbing fastness in the presence of the mordant.

5.1.2.5 Functional properties of turmeric related to textile application

Turmeric also has antibacterial and anti-inflammatory effects. Natural colorants extracted from turmeric exhibited excellent antimicrobial activities and related wound healing properties [17]. Silk fabrics dyed with an extract from *C. longa* rhizome using copper sulphate, ferrous sulphate, and potassium aluminum sulphate as pre-mordants possessed desirable antibacterial properties and 3% (owf) copper sulphate giving complete antibacterial activity against *Staphylococcus aureus* (Gram-positive) and *Escherichia coli* (Gram-negative) [11]. The study also indicated that an increase in the dye concentration leads to a more efficient antibacterial activity and 30% (owf) of turmeric gave the optimum level of antibacterial activity. Nylon fabric dyed with various concentrations of turmeric extract using different metallic mordants displayed excellent antibacterial activity in the presence of ferric sulphate, cupric sulphate, and potassium aluminum sulphate, and exhibited good and durable fastness properties [13]. Cotton yarns colored with turmeric and coated with chitosan provide high antibacterial action against bacteria (*E.coli* and *S.aureus*). Also, the yarn coated with chitosan dyed to a darker shade compared to uncoated yarn for the amount of the dye used [18]. Colorant from turmeric can have UV protection properties and can block almost 100% UV-rays when used to dye polyester. On coating the fabric with chitosan there was no change in UV protection property though the slight change in the shade was noted [19].

5.1.2.6 Case study-1

Turmeric (*C. longa L.*) extract was used to dye cotton using bio-mordants (*Citrus limon* and *Colocasia esculenta*), and for comparative purposes, metallic mordants (potassium dichromate and potash alum) were also used [20]. The samples were pre-mordanted (soaked) in the bio-mordant extract for different durations before dyeing; for the metallic mordants, they were boiled with the mordant solution at 80°C for 50 min followed by cooling for 60 min in the solution itself. The effect of mordanting time on the color strength was evaluated for the bio-mordants. The surface color strength (K/S) of bio-mordants (*Citrus limon* and *Colocasia esculenta*) pre-mordanted cotton increased with an increase in mordanting time for both the bio-mordants (**Table 2**). Cotton pre-mordanted with lemon containing significant amounts of tannins showed the highest surface color strength (K/S) among all mordants used. The effect of moisture absorption on the hue of the dyed fabric was also studied. For this, the dyed specimens were stretched and tied over the mouth of steel tubes containing 100 ml water each. The specimens were

Treatments	Mordanting time		
	1 hour	3 hours	5 hours
Unmordanted sample	4.0		
Cotton pre-mordanted with Colocasia	4.1	4.5	5.1
Cotton pre-mordanted with Lemon	7.0	7.3	8.6
Cotton pre-mordanted with potassium dichromate	7.5*	—	
Cotton pre-mordanted with potash alum	4.0*	—	—

**for 1 hr. 50 min*

Table 2.
Surface color strength of cotton dyed with turmeric pre-mordanted with different mordants for different time duration.

maintained under normal conditions of 25°C and 70% relative humidity for 24 h after which the face and rear side of the specimens were visually observed for any change in hue. Under acidic conditions (below pH 4), curcumin gave a yellow appearance, in alkaline pH, it changed its hue as the dyed cotton specimen absorbed moisture changing its pH and thus showing a significant change in hue on both side of the fabric. Furthermore, the visual uniformity of the dyed samples was found to be excellent for both bio-mordants. Due to the presence of citric acid, turmeric gave uniform color in low acidic conditions (around pH 4); at higher pH (pH 4 to 5) it showed a reddish color. Color fastness to rub (dry and wet), water (EN ISO 105 E01–2013), wash (ISO 105 C06), and perspiration (EN ISO 105 E04–2013) were found to be superior for the bio-mordanted cotton and the values ranged from 3 to 4–5 in most all cases.

5.1.2.7 Case study-2

Aqueous extract of turmeric was used to dye cotton fabric using aluminum sulphate as a mordant [15]. The effect of different mordanting techniques (per, post, and simultaneous) on the surface color strength of the fabric was evaluated (**Table 3**). Simultaneous dyeing and mordanting sequence gave maximum dye uptake probably due to the mordanting of cotton with aluminum sulphate mordant and formation of a complex between the color component of the dye curcumin and the metal mordant. Also, turmeric being a direct type of dye exhausted well in the presence of a salt-like alumnium sulphate (mordant) and hence simultaneous mordanting sequences gives better results (K/S).

New and uncommon compound shades were developed through combination dyeing of the cotton combination of turmeric (yellow dye) with using madder (red dye), and turmeric (yellow dye) with red sandalwood (red dye) in different

Mordanting Technique	K/S at λ_{max} (450 nm)
Pre	0.4
Post	0.3
Simultaneous	1.5

Table 3.
Surface color strength (K/S) of cotton dyed with aqueous extract of turmeric using aluminum sulphate as a mordant by the different mordanting sequences.

Dye	Amount of dye when used singly					Proportional ratio of the dye in the mixture	Calculated value for the combined shade	Observed value for the combined shade
	100	75	50	25	0			
Turmeric	—	0.7	0.3	0.2	—	100:0	—	1.5
Madder	—	0.3	0.3	0.4	—	75:25	0.7 + 0.2 = 0.9	1.0
						50:50	0.3 + 0.3 = 0.6	0.9
						25:75	0.2 + 0.4 = 0.6	0.8
						0:100	—	0.6
Turmeric	—	0.7	0.3	0.2	—	100:0	—	1.5
Red sandalwood	—	0.3	0.3	0.2	—	75:25	0.7 + 0.2 = 0.9	0.9
						50:50	0.3 + 0.3 = 0.6	0.7
						25:75	0.2 + 0.3 = 0.5	0.6
						0:100	—	0.4

Table 4.
Surface color strength (K/S) of cotton dyed with a mixture of dyes (turmeric with madder and turmeric with red sandalwood) in different proportion by the simultaneous mordanting and dyeing sequence using aluminum sulphate as a mordant.

proportions by the different mordanting and dyeing process. A synergistic effect in the color interaction between the observed and calculated K/S values (calculated values were derived by adding the individual K/S value of the respective proportion of the two dye components on the fabric) was observed; the observed K/S values of the dyed cotton samples were always higher than the calculated or expected K/S values indicating the color value of the mixed dye system to be always higher. Also, an increased amount of turmeric in the mixture increased the dye uptake (K/S) values (**Table 4**).

5.1.3 Application of turmeric in food coloration

Curcumin is a polyphenol found naturally in turmeric rhizome that has antiinflammatory, antioxidant, anticancer, and immunosuppressive activities. It is used mainly in the development of dairy products as the presence of fat (triglycerides) enhances the solubility of curcumin [21]. While few studies have been carried out on colouration of food using turmeric, most of them focus on its functional aspects. Improvement in the sensory attribute and antioxidant potential of ghee has been reported by the addition of 160–350 ppm of curcumin [22]. The turmeric powder improved the oxidative stability and microbiological quality of soft cheese [23]. Turmeric extract rich in curcumin reduced the aging of fresh lamb sausages during modified atmospheric packaging by causing less generation of related volatile compounds due to its antioxidant capacity [24]. The addition of turmeric to the dough of biscuits and breads greatly improved the antioxidant potential and organoleptic properties of breads and biscuits [25].

5.1.3.1 Color produced

Turmeric when applied to food yields a bright orangish-yellow shade.

5.1.3.2 Extraction & application conditions

Curcumin is mainly dissolves in oils and alcohols. It is not stable at alkaline conditions especially at pH above 7.5 though it is quite stable in temperatures generally used for processing foods. Curcumin is complexed with aluminum ions as it is light sensitive.

5.1.3.3 Functional properties related to food application

Curcuminoids present in turmeric possesses anti analgesic, anticarcinogenic, antiinflammatory antioxidant, antiseptic properties. It also helps in the prevention, palliation, or treatment of various disorders such as diabetes, cholelithiasis, diabetes mellitus, foodborne illnesses, and circulatory disorders [26–28]. Moreover, it also acts as a potent food preservative as it slows down lipid oxidation and possesses antimicrobial activity.

5.1.3.4 Case study-3

The effect of heat treatment and conventional sun drying on the color of fresh turmeric rhizome was evaluated in terms of its hue, yellowness, and brightness (L*, a*, and b* color coordinates) [29]. Turmeric rhizomes were subjected to heat treatment at varying temperatures (50–100°C) for different time periods (10–60 minutes). The rhizomes were cooked at 100°C and then sun-dried for 15 days. The rhizomes were brightened (L*) and yellowed (b*) after being heated at 60-80° C. Heat treatment from 60 to 80°C increased the brightness (L*) and yellowness (b*) of the rhizomes; the values remained the same and did not change with further increase in temperature. The phenolic activity of oxidases in turmeric decreased with an increase in temperature and this led to a decrease in browning of the sample while inversely increasing its hue to a yellower shade and brightness. Though the heat treatment did not significantly decrease the concentration of curcuminoids, sun drying caused a significant reduction in curcuminoids (4–5%). Heat treatment thus enhanced the color of turmeric and maximum brightness was observed at 80°C for 30 minutes.

5.1.3.5 Case study-4

The impact of irradiation on the color stability of curcuminoids was examined and curcumin reagent (curcumin, DMC, and BMC; 79.4, 16.8, and 3.8% - w/w) was irradiated with fluorescent light (27 watt) for 24 hours using a household fluorescent lamp [30]. The color intensity was analyzed by measuring absorbance at 435 nm and curcuminoids before and after treatment were quantified using HPLC. Turmeric pigments (oleoresin and curcumin) were not stable under light, and their photo-degradation was lower when present in higher concentrations. An increase in concentrations of the sample (20–1000 μg/mL) resulted in a loss in color intensity of both oleoresin and curcuminoids in turmeric (**Table 5**).

5.2 Annatto

Bixa orellana is a perennial, tall shrub bearing bright white or pink flowers and red-brown fruits in the form of globular ovoid capsules or seed pods with delicate spines. The pods are grouped in clusters and each contains 30–45 cone-shaped seeds covered by a pericarp rich in the red-orange pigment, annatto. *B. orellana* is native

Concentration (μg/mL) of the sample	Color intensity	
	Turmeric oleoresin	Curcuminoids
20	65.4%	63.0%
200	38.9%	46.2%
1000	28.6%	27.0%

Table 5.
Loss in color intensity of different pigments (i.e. oleoresin and curcuminoids) in turmeric due to light irradiation.

and grows wild in northern South America and Central America. Later in the 16th and 17th centuries, *B. orellana* was distributed to the Caribbean, Hawaii, and South-Eastern North America, Southeast Asia, and Africa. It is cultivated primarily for its red seeds in India, Sri Lanka, and Java. In India, *B. orellana* is cultivated for its seed across Orissa, Andhra Pradesh, and Maharashtra. 70% of the world's coloring agents derived from natural sources come from annatto [31]. Its color is used in food, textile, paint, and cosmetic industries. Also called achiote or bijol it is used as a natural orange-red condiment/spice in the food industry and is used in the bleaching of dairy food products. It is soluble in lipids and is therefore used for imparting red to orange-yellow color to processed food. Annatto is also known as lipstick tree [32] and is used in cosmetics for the production of sunscreens [33], nail gloss, hair oil, and soap. Its medicinal value is associated with its antibacterial, antifungal, antioxidant, antibiotic, and antiinflammatory properties. It has shown anticancer, enhanced gastrointestinal motility, neuropharmacological, anticonvulsant, analgesic, and antidiarrheal activities and has been used as a laxative, cardiotonic, and expectorant, and for wound healing purposes. The dye is also used in the printing and dyeing of textiles like cotton, wool, and silk.

Genus: *Bixa* | **Species**: *orellana* | **Family**: Bixaceae.

Common name: Achoite | **Local name**: *Latkan or sinduri.*

Part of the plant used for coloring: Seeds.

5.2.1 Coloring pigment/component

Of the total carotenoid pigments present in annatto, 80% consists of the red pigment, bixin, and a yellow pigment, norbixin or orelline. Bixin is a yellowish-orange-red dye that is high in carotenoid pigments and is derived from the thin seed coat of *B. orellana* seeds. Bixin occurs in nature as monomethyl ester of the dicarboxylic carotenoid compound [6,6′-diapo-ψ-ψ′-carotenedioic acid monomethyl ester] i.e. 16-*Z* (*cis*) form, but during extraction, it isomerizes to its 16-*E* (*trans*) form called isobixin. Norbixin is a naturally occurring demethylated derivative of bixin used for commercial purposes. Besides bixin and norbixin, other compounds such as beta-carotene, cryptoxanthin, lutein, zeaxanthin, orellin, bixein, bixol, crocetin, ishwarane, ellagic acid, salicylic acid, threonine, tomentosic acid, tryptophan, and phenylalanine are also found in the seeds of annatto. Bixin belongs to the direct/acid dye class [34]. Bixin content influences the color value of the annatto extract. With higher amounts of bixin, the L^* and b^* values decreased (darker and yellower) whereas a^* values increased (redder) under the Hunter measurement scale [35]. For Lovibond values, for the same dyes, the R-values increased with the increase in the concentration of bixin while the Y values remained the same. The low purity dye (CFTRI method) showed a higher b^*/a^* values as compared to the high purity dye (new patented process by CFTRI), whereas a reverse trend was

Concentration of bixin in mg/L extracted by the patented method	Hunter		Lovibond	
	L*	b*/a*	Y	R
10	15.2	2.3	40	6.0
20	14.1	1.9	40	8.0
30	12.9	1.5	40	9.0
40	12.4	1.3	40	10.0
50	12.2	1.2	40	11.0
100	9.7	0.8	40	17.0

Table 6.
Effect of bixin concentration on color values (hunter and Lovibond) [35].

observed with respect to the Y/R values. However, a* and R-values which corresponded to red color increased with an increase in concentration in both color measuring systems irrespective of dye purity [35]. With annatto giving orange shades (combination of yellow and red) b*/a* (degree of yellowness) values were also assessed. With an increase in the bixin concentration, the b*/a* decreased indicating a more yellow color. The study also indicated that the Lovibond color was more influenced by the source of dye and its purity as compared to the Hunter values (**Table 6**).

5.2.2 Application of annatto in textile colouration

Natural fibres like cotton [34, 36], silk [37] and wool [32] and also synthetic fibers like nylon and polyester [38] have been dyed with *B. orellana*. Leather has been dyed in the bright red shade with excellent rub fastness using bixa extract [39].

5.2.2.1 Color produced

Yellow and orange can be produced from *B. orellana*. Though annatto seed extract gives an orange-red color, the hue depends on the solvent used for extraction [40]. The wavelength of maximum absorption for annatto is 458 nm [41].

5.2.2.2 Extraction

Commercial preparations consist of solutions or suspensions of the pigment in vegetable oil or as a water-soluble form in dilute alkaline solution. Content of total phenols (TP) increases with an increase in pH and higher TP contents were obtained at an extraction time of 60 h and a solvent/seed ratio of 4 ml/g of the extract [42]. The primary pigment *cis*-bixin is partially transformed to the *trans* isomer and a degradation product when heated [43]. Microwave-assisted extraction using ethyl acetate solvent also gives good pigment yield [44]. Though the total dye yield (**Table 7**) on the extraction of annatto seeds by the new patented process by CFTRI, Mysore, 2004 was less than the dual solvent extraction method (CFTRI method), the more purer patented process gave higher bixin and nobixin yields (g/100 g) [35].

5.2.2.3 Dyeing conditions

B. orellana gives beautiful shades on cotton in alkaline medium using inorganic salts as mordants [45]. Woolen yarns can be dyed with bixa extract in acidic,

Extraction condition	Dye yield (g/100 g)	Bixin (g/100 g)	Norbixin (g/100 g)
Bixin/norbixin dye from Indian seeds by CFTRI method	2.3	21.9	18.5
Low bixin/norbixin dye from Indian seeds by the special patented method	2.0	13.9	12.4
High bixin/norbixin dye from Indian seeds by special patented method CFTRI method	1.0	60.2	55.4

Table 7.
Total yield of dye with bixin and norbixin content in Indian seeds of annatto extracted by different processes [35].

neutral, and alkaline media using ferrous sulphate, stannous chloride and alum as mordants. Regardless of the presence or absence of mordants, dyeing silk and wool fabric with an aqueous extract of annatto seeds is best successful at pH 4 [41].

5.2.2.4 Fastness

B. orellana reportedly has moderate to poor light fastness, but moderate to excellent fastness to washing, rubbing, and perspiration.

5.2.2.5 Functional properties of annatto related to textiles application

Extract of annatto has remarkable antimicrobial and antioxidant properties and a study revealed that the annatto dye had a bactericidal effect and could reduce *E. coli* activity [46]. Ethanolic extract form seeds showed broad spectrum antibacterial activity against *Bacillus subtilis*, *Staphylococcus aureus*, *Streptococcus pyogenes*, *Salmonella typhi*, *Pseudomonas aeruginos*a, *Escherichia coli*, and *Candida albicans* [47]. When compared to the traditional method, annatto dye extracted by the ultrasound aided technique has higher antibacterial and antioxidant activities [48]. Gram-positive bacteria show more sensitivity to annatto dye extracted by conventional and ultrasound-assisted extraction methods than gram-negative bacteria and. *B. subtilis* showed the lowest sensitivity toward annatto dye, while *Escherichia coli* gave the highest sensitivity. Annatto dye extracted by UAE showed a bactericidal effect against *Salmonella enteritidis* [48]. Annatto extracted from annatto seeds (*Bixa Orellana*) using ultrasound technique was used to color biodegradable films based on poly (3-hydroxybutyrate). Photo-degradation under UVA exposure of colored films showed improvement as suggested by the SEM micrographs, and the film also showed good thermal stability as confirmed by the thermo gravimetric analysis [49].

5.2.2.6 Case study-5

Cotton, wool, and silk were dyed with an aqueous extract of the *B. orellana* seed powder using the one-step simultaneous sequence and the two-step pre-mordanting sequence by the ultrasound technique [50]. Enzymes were used along with tannic acid to treat the fabric; cellulose and amylase for cotton and protease for silk and wool. The exhaustion of color on the fiber ranged from 55 to 63% for the ultrasound technique, while it was lower ranging between 42 and 46% for the conventional exhaust dyeing procedure for all three fibers. Because bixin and norbixin are polar and acidic, they have a stronger affinity for wool and silk (protein fibers) which have more polar groups than cotton (cellulosic fiber). Enzyme treatment increased

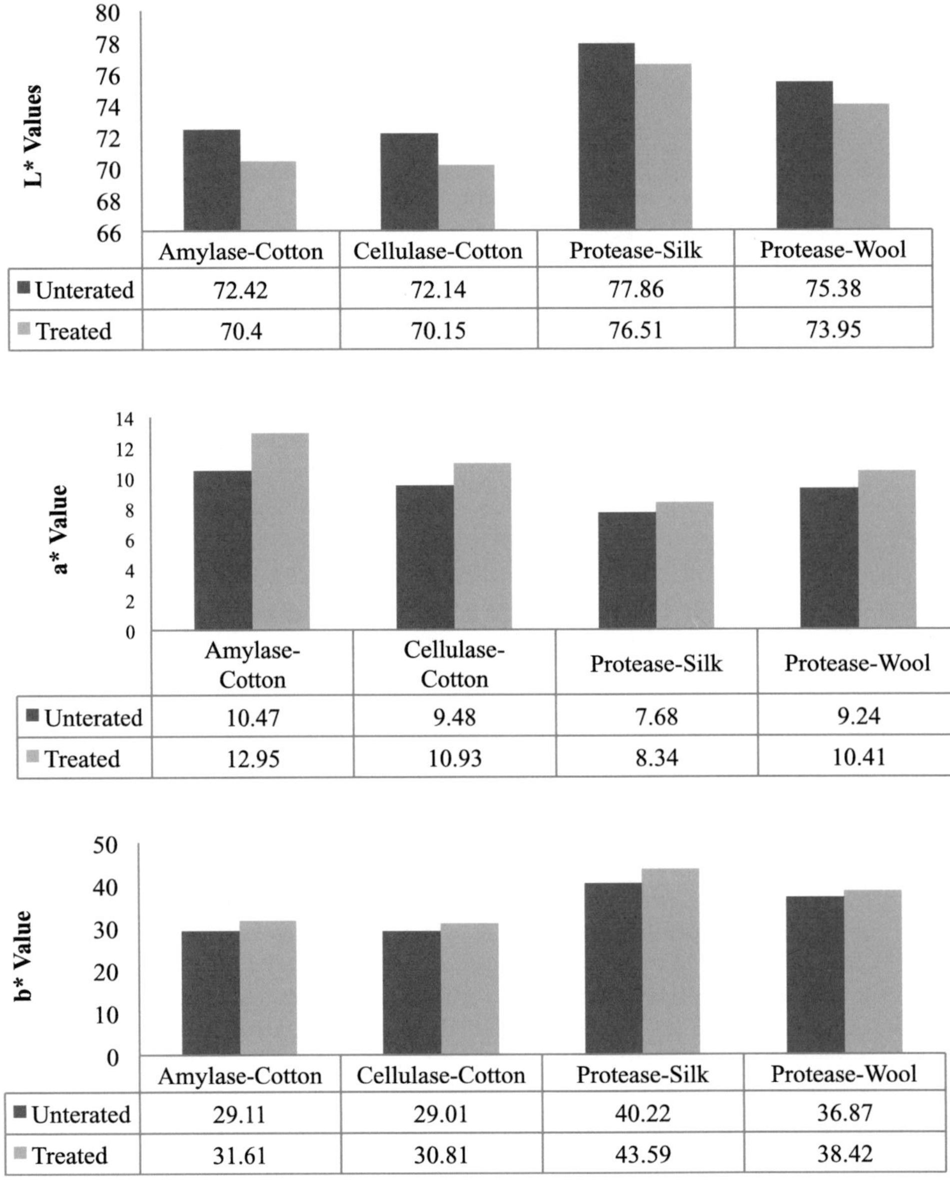

Figure 1.
Effect of enzyme treatment on the color related properties (L, a* and b*) of cotton, wool, and silk were dyed with an aqueous extract of the* Bixa orellana *seeds.*

the surface color strength (K/S) for all the fibers and lowered the L* values (indicating darker shades). It also slightly increased a* and b* values of all the dyed fibers indicating redder and yellower i.e. probably an orangish tone (**Figure 1**).

5.2.2.7 Case study-6

Eco-friendly bamboo fiber was dyed with *B. orellana* using potash alum as mordant. The dyeing procedures variables of time, temperature, and pH were optimisation for the pre-mordanting process. An increase in time from 15 min to 60 min increased the color yield and a sharp liner increase in the surface color depth (K/S) was also observed with an increase in temperature from ambient temperature

Varying Parameters		K/S at λ_{max}	L*	a*	b*
Control (desized and potash alum pre-mordanted bamboo)		0.1	89.3	−0.2	6.6
Variation in time (in min)	15	4.4	−23.6	30.7	1.2
	30	4.5	−24.3	29.8	1.2
	60	4.7	−24.2	30.0	39.6
Variation in temperature (°C)	Ambient	3.9	−19.9	28.2	42.5
	60	4.7	−23.3	29.8	42.2
	80	5.6	−25.8	30.8	42.5
Variation in pH	2	1.4	−15.3	17.3	25.8
	7	3.8	−23.0	28.0	37.0
	10	6.5	−26.5	33.0	45.8

L – lightness/darkness; a* – greenness/redness difference; b* – blueness/yellowness; and CDI – color difference index*

Table 8.
*K/S (color strength) and other color related parameters of bamboo fabric pre-mordanted with potash alum and dyed with aqueous extract of annatto seeds (*Bixa orellana*) under variable conditions of dyeing.*

to 80°C. The K/S value of the mordanted bamboo fabric was significantly higher when dyeing was carried out under alkaline conditions. The optimum conditions of dyeing of the potash-alum pre-mordanted bamboo fabric with aqueous extract of seeds of annatto (*B. orellana*) was thus reported as dyeing time −60 min, dyeing temperature −80°C, and pH–12. Varied shades of orange (sherbet-orange to ginger-orange and apricot colors) were obtained which were dark as indicated by negative L* values. The shades were redder as indicated by positive a* values and yellower as reflected by positive b* values (**Table 8**). The yellower tint (b*) was more pronounced compared to the redness tint (a*) in most cases [51].

5.2.3 Application of turmeric in food coloration

Annatto (E-160B) is a natural yellow-orange dye obtained from *B. orellana*, with less toxicity. Compared to synthetic food colorants, it typically demonstrates superior biodegradability and compatibility with the environment [52]. Annatto colorants are used extensively in the food industry, particularly in the processing of dairy and meat products.

5.2.3.1 Color produced

Annatto gives a yellow to orange-red shade on food.

5.2.3.2 Extraction & Application conditions

Annatto is water-soluble and can be mixed with sugar powder or potassium carbonate. The pigment is not heated stable. Moreover, there is a considerable loss of pigment due to deep-fat frying at high temperatures (> 200°C). It is stable at a pH 5.0–10.

5.2.3.3 Functional properties related to food application

Extract of annatto seed possess antimicrobial properties and decrease the growth and activity of *B. subtilis, Staphylococcus aureus, Streptococcus Clostridium thermophilus, perfringens, Lactobacillus casei, Lactococcuslactis, Candida albicans, Candida famata, Rodotorula* species, *Aspergillus* species, and *Neurosporacrassa*. The seed extract showed strong inhibition of triglycerides oxidation in rapeseed oil [53] and norbixin was able to inhibit oxidative deterioration of olive oil [18]. Annatto seeds' antioxidant capabilities aid in the treatment of cardiovascular disorders and have been found to decrease the oxidation of low-density lipoprotein in vitro (LDL). Extracts can protect DNA from oxidative damage thereby controlling serious consequences in some age-related cancers.

5.2.3.4 Case study-7

Color from annatto seeds is safe for human consumption compared to the synthetic colorants commonly used in sweetmeats. *Jilebi* and *jangri* are popular Indian sweetmeats colored in various shades of yellow and red. Bixin extracted from annatto seeds by a dual solvent method (washing the seeds with a non-polar solvent like hexane and then recovery of the dye in acetone) was used to color *jilebi* and *jangri* and its effectiveness was compared to the commercially available counterparts that were colored using synthetic food colorants [54]. The dye was converted to norbixin using potassium hydroxide (alkaline pH) and diluted with potassiun carbonate (K_2CO_3) to get its water-soluble form. Two water-soluble formulations of the colorant (bixin), one having potassium carbonate (with norbixin content, 10.6%) and the other with sugar (with norbixin content, 11.24%) were prepared. The first formulation with potassium carbonate was used to prepare the batter for the sweetmeat, while the other sugar formulation was used for further sweetening the sweetmeat after frying. The bixin extracts were applied in varying concentrations in both the formulations for the preparation of *jilebi* and *jangri*, and the effect in terms of red and yellow color units using Lovibond tintomer was evaluated. The effect of frying temperatures (142°C and 172°C) on the color value of both *jilebi* and *jangri* were also evaluated. With the increase in the norbixin concentrations, the R values indicated an increasing trend; while the Y values were higher and constant. Commercial *jalebis*' R values matched the R and Y values of an annatto solution comprising containing 2.5 and 5 mg/kg of norbixin for sugar-based formulations; while potassium carbonate-based formulations, 5 mg/kg of norbixin gave comparable results (R values) though Y values were higher (**Table 9**). For *jangris*, 40 mg/kg of norbixin gave comparable R and Y values for

Sample	Red values	Yellow values
Commercial ***jalebi***	1.7–4.1	9.0–20.0
2.5 mg/kg of nor-bixin (in sugar based formulation)	2.0	20.0
5 mg/kg of nor-bixin (in sugar based formulation)	3.0	20.0
5 mg/kg of nor-bixin (in potassium carbonate based formulation)	3.1	40.0
Commercial ***jangri***	9.1–10.8	20.0–20.7
40 mg/kg of nor-bixin (in sugar based formulation)	10.0	30.0
40 mg/kg of nor-bixin (in potassium carbonate based formulation)	10.0	40.0

Table 9.
Tinctometer color values of commercial jalebis *and* jangris, *and of the solutions containing optimized concentrations of norbixin with potassium carbonate and sugar.*

both sugar and potassium carbonate-based formulations. Soluble annatto dye sugar powder formulation gave much better results with lower concentrations of dye as compared to soluble annatto dye potassium carbonate formulation. Water-soluble annatto dye sugar solutions at concentrations 5 mg/kg were found to be optimum for *jalebi* and 30 mg/kg for *jangri* as the colored developed was similar that those found in the commercial counterparts. Excellent color matching was observed and no difference in redness and yellowness in the color of the product were reported due to the effect of high temperature used in frying.

5.2.3.5 Case study-8

The solubility of bixin in oil and norbixin in water determines its usage. Annatto dye formulations suitable for dairy products like cheese and butter were developed and compared to their commercially available counterparts [55]. Three formulations were prepared; water-soluble solution using K_2CO_3, oil-soluble formulation using vegetable oil, and oil/water-soluble formulation using propylene glycol solution. The formulations were applied at different concentrations in cheese and butter. Lovibond Tintometer was used to measure the color of the commercial and experimental samples. Annatto dye oil/water soluble propylene glycol formulation was found to be the most effective formulation for imparting yellow color with good brightness to various dairy products (**Table 10**). Butter containing 3.75 mg/kg and 5 mg/kg of oil/water propylene glycol formulation closely resembled the commercial butter samples made using synthetic dyes. In the case of cheese, creamy yellow shade imparted by oil/water propylene glycol formulation at a concentration of 3.75 mg/kg looked very similar to the color of the commercial cheese sample.

Sample	**Concentration (mg/kg)**	**R values**	**Y values**
Commercial butter	—	1.2 ± 0.26	4.0 ± 0.36
Butter with oil soluble annatto extract formulation	3.8	0.8 ± 0.17	2.0 ± 0.26
	5.0	1.0 ± 0.17	3.0 ± 0.26
Butter with water soluble annatto extract formulation	3.8	0.9 ± 0.10	2.5 ± 0.26
	5.0	1.0 ± 0.10	3.3 ± 0.20
Butter with oil/water annatto extract formulation	3.8	1.1 ± 0.17	4.0 ± 0.46
	5.0	1.5 ± 0.26	6.0 ± 0.36
Commercial cheese	—	1.6 ± 0.26	4.6 ± 0.26
Cheese with oil soluble annatto extract formulation	3.8	1.1 ± 0.10	4.0 ± 0.17
	5.0	1.3 ± 0.26	3.3 ± 0.20
Cheese with water soluble annatto extract formulation	3.8	1.2 ± 0.18	2.5 ± 0.26
	5.0	1.2 ± 0.20	3.0 ± 0.30
Cheese with oil/water annatto extract formulation	3.8	1.4 ± 0.12	4.6 ± 0.21
	5.0	1.8 ± 0.10	5.0 ± 0.21

Table 10.
Lovibond tintometer readings of commercial and experimental test samples of butter and cheese.

5.3 Cochineal

Cochineal is a natural dye made from the pulverized and dried corpses of a female sessile parasite found in tropical and subtropical South America and North America. Dyeing of cochineal extract is mainly practiced in Mexico and Peru. Cochineal extracts have been used over ages as colorant for food, textiles, cosmetics, pharmaceuticals, and plastic applications.

The dye has mostly been used in the dyeing of silk, wool, cotton, and natural pigments (lakes) obtained from cochineal insects were used for paintings, frescoes, and restoration processes [56]. It's the only natural red color that's been allowed by the FDA for use in food and cosmetics, and it's frequently used as a substitute for the infamous Red Dye #2.

Genus: Dactylopius | **Species**: coccus | **Family**: Dactylopiidae.

Common name: Cochineals | **Local name**: *Cochineal keet.*

Part of the plant used for coloring: Cochineal insects are found on the pads of prickly pear cacti in the genus *Opuntia* or *Nopalea* and are collected by brushing them off the plants, killed (by immersing in hot water or exposure to sunlight, steam, or dry heat of an oven), dried and powdered to get the dye. One pound of cochineal nectar requires 70,000 insects.

5.3.1 Coloring pigment/component

The important color producing components in cochineal extract are carminic acid, kermesic acid and flavokermesic acid [57–59]. Cochineal's coloring ability is due to cochinealin, or carminic acid (80–86%) with anthraquinone as the chromophore and –COOH, –OH, >C=O, and $-CH_3$ as auxochromes. The bodies of female insects contain up to 25% of their dry weight of this pigment. Glyceryl myristate (a lipid) and coccerin (cochineal) are also found in cochineal. Carmine is formed by precipitating carminic acid onto an alumina hydrate substrate and dried to typically 50 percent concentration. Carmine is insoluble in water but is water-soluble when treated with a strong alkali [60]. Carminic acid showed a moderately strong correlation with chromatic values (a*) from the pigment extract (**Table 11**). Also, there were no significant differences in the tint value of the samples containing different proportions of carminic acid [61].

5.3.2 Application of annatto in textile coloration

Cochineal was considered as one of the great treasures of the New World in the 16th–18th centuries, and along with alkanet, madder, kermes, and lac it formed a

Carminic acid content (percent) in different types of turmeric	L	a*	Tint (A420/A500) (ratio between yellow and red pigment)
12.8	19.5	3.9	0.44
15.8	19.4	3.8	0.44
16.0	19.1	3.7	0.45
17.9	19.4	3.6	0.46
19.7	19.4	3.3	0.44

Table 11.
Variation in color values with respect to changing carminic acid n content in turmeric taken from difference different geographical origin.

source of natural red dye for textiles. Cochineal dyed textile fibers in intense red colors with excellent fastness and was the dyed textiles were highly prized. There are several studies on the use of cochineal for dyeing different fibers; cotton has been dyed with cochineal [62, 63] as also wool [4] and silk [64]. Cochineal extract was used to dye silk and wool by the simultaneous dyeing and mordanting process using 1 gpl and 5 gpl of the dye and 1.5 gpl potash alum and copper sulphate as mordants at pH 4 and 80°C for 90 minutes using liquor ratio 1:40 [65]. Polyamide fabric has been successfully dyed in a range of shades with cochineal using different mordants and mordanting methods [66].

5.3.2.1 Color produced

Cochineal produces scarlet, crimson, orange, and other range of fuchsias, reds, and purples on textiles. Different mordants produce different shades; blue-red/-reddish-purple color with alum, maroon-red with copper, purple with iron. The addition of cream of tartar into the dye bath during the dye process will shift the color from a reddish-purple to a vivid flag red color. A combination of mordants also produces different colors like rich red when tin and alum are combined, purple-red when alum and iron are combined, and fuchsia to red shades with a combination of alum and cream of tartar. Over dyeing of cochineal with madder gives a good red, whilst cochineal over-dyed with indigo yields a range of light-fast violets and purples. Cochineal carmin has a maximum absorption wavelength (max) of 520 nm [67]. When carmin is esterified, the hydroxyl groups transform to carbonyl groups, lowering the electron cloud density and resulting in light shading effects [68].

5.3.2.2 Extraction

The bodies of the insect, *Dactylopius coccus* contain 19–22 percent carminic acid, which can produce crimson and scarlet colors. To preserve the dye without rotting, the insects are dried to roughly 30 percent of their original body weight. The female cochineal insects are processed by immersing them in hot water or exposing them to sunshine, steam, or the dry heat of an oven to extract carminic acid. Each process generates a different color. The dried and powdered insect corpses are cooked in ammonia or sodium carbonate solution, the insoluble debris is removed by filtration, and alum is added to the clear salt solution to precipitate the red aluminum salt to make carmine, a more pure version of cochineal. Colorant extracted from cochineal in acid solubilized medium enhance the color characteristics of bio-mordanted silk fabric [64].

5.3.2.3 Dyeing conditions

pH of dye-bath has a great influence on shades obtained with cochineal though they do not impact the fastness properties of the dyed textiles.

Since the phenolic groups in cochineal are acidic, carminic acid is pale orange in low pH, but it changes to red in slightly acidic and neutral pH, and finally turns violet in alkaline solution [69]. Alkaline medium is favorable for dyeing cotton fabrics with cochineal extract and pre-mordanting cotton with alum and tannic acid mordant mixture improves the color yield [63]. Carminic acid also forms complexes with several metals ions, which act as acceptors to electron donors to form co-ordinate bonds with water-insoluble dye molecules. This complex formation between the dye and the mordant shifts the maximum absorption in the visible range to higher wavelengths with an apparent increase in color intensity. Tin-based mordanting gives a brighter, but higher lightness (L^*) value on wool dyed with cochineal than other mordants [70]. The pre-mordanting method is preferred for aluminum and chromium salts, while the post-mordanting method is preferred for copper, tin, and iron salts in order to improve the color yield of wool dyed with cochineal extracts [71]. Catonization of cotton fabric [72] or its treatment with chitosan [70, 73] increases the color value of the cochineal dyed fabric. The optimum dyeing conditions for dyeing cotton with cochineal has been identified as temperature −60°C, time −60 min, MLR–1:40 liquor ratio [74].

5.3.2.4 Fastness

Cochineal generally dyes textiles with excellent light and wash fastness. It gave moderate to good fastness properties on cotton [74] and moderate (grade 3) to very good (grade 4–5) washing fastness, and moderate (grade 5) to excellent (grade 7–8) light fastness on wool yarns [75]. Excellent fastness properties have also been reported on wool dyed with cochineal under the influence of microwave treatment and bio-mordants like heena and pomegranate [4].

5.3.2.5 Functional properties of cochineal related to textiles application

Cochineal imparted antibacterial property to wool, silk, nylon, cotton, and viscose rayon fabrics [71, 76, 77]. Nylon yarn dyed with cochineal dye showed limited antibacterial activity, which increased on mordanting with copper and tin [76]. Excellent UV protection properties (UPF > 100) were observed on wool dyed with cochineal and this was higher for copper sulpate mordant compared to alum and also improved with the increase in dye concentration [65]. UPF values for silk dyed with cochineal was less than 50 at lower concentrations of the dye, but it was very good and in the acceptable range (UPF > 50) with a higher concentration of the dye and in the presence of copper sulphate mordant [65].

5.3.2.6 Case study-9

Woolen yarns were dyed with an aqueous extract of cochineal in presence of five different mordants (aluminum sulphate, stannous chloride, ferrous sulphate, citric acid, and cream of tartar i.e. potassium hydrogen tartarate), singly and in combination, using the pre-mordanting method as well as simultaneous mordanting methods [75]. During dyeing, the carbonyl group (>C=O) and alpha hydroxyl groups (–OH) in the anthraquinone moiety of carminic acid/kermesic acid of cochineal forms a coordinate complex with the metal cation of the mordant. The carboxylic acid group of the cochineal dye can also tautomerize and easily ionize into carboxylate anion ($-COO^-$) forming ionic bonding with $-NH3^+$ group of the wool fiber. In this way, metal-dye-fiber coordination complexes are formed between the mordant, dye, and the fiber. The anthraquinone-metal combination formed by cochineal and the metal mordant causes a red and blue shift in the visible region, i.e. between 460 and 570 nm, resulting in scarlet-red to purple colors [78]. Due to the H-substitution of the hydroxyl group bonded to C5 of the dye molecule by each metallic ligand, carminic acid present in the cochineal dye induces a bathochromic shift of the main hue to red when it interacts with metal cations during mordanting [79]. This happens when the bonding occurs between the 2-hydroxy group of dye molecule and metal cation [80]. But if bonding between dye and metal ion occurs in 7-hydroxy group, the complex could induce a small blue shift [80]. The bluish-purple color was obtained on unmordanted wool and a range of colors from scarlet-red to black on mordanting with the various mordants. In the case when mordants were used in combination, the final color depended on the chelating property of the dominant mordant, which forms more coordination complexes with the cochineal dye than the other mordants. Thus, ferrous mordant combinations gave grayish chrome; stannous mordant combinations gave reddish chrome and aluminum mordant combinations gave purple chrome. The redness/greenness (a* values) values of dyed samples from both the pre-mordanting method and simultaneous mordanting procedures were positive, indicating that all colors obtained using cochineal dye were in the red-purplish range. All dyed samples irrespective of the mordanting procedures showed an increase in yellowness (b* values) after mordanting and consequently, the color of dyed samples shifted from bluish (higher negative b* values) to yellowish (lower negative or positive b* values). In the pre-mordanting method, the metal cation of the mordant probably diffused well inside the fiber matrix-forming ionic bonding with functional groups of wool fiber before dyeing. During this dyeing process, this metal cation fixed on the fiber probably formed coordinate bonding with the cochineal dye molecule resulting in more aggregation of the dye molecules with the metal cation and formation of dye-fiber-metal complex inside the fiber. Contrarily in the simultaneous dyeing and mordanting method, the coordinate complex between the metal cation and the cochineal dye molecule was probably formed both in the dye-bath as well as inside the fiber matrix leading to lesser aggregation of dye-metal complex inside the wool fiber. Thus darker shades were obtained by the pre-mordanting process and the lightness (L*) of dyed was found to be higher in the case of simultaneous dyeing and mordanting process.

5.3.2.7 Case study-10

Wool was dyed in purple shades with cochineal and metal mordant (aluminum sulphate) and bio-mordant (chitosan) using the pre-mordanting process [81]. Results show that K/S value of wool mordanted with chitosan was higher than when mordanted with aluminum sulphate. Dye uptake increased with an increase in the concentration of the bio-mordant but beyond 1000 mg/L concentration, the K/S decreased. The decrease in dye absorption at higher bio-mordant concentrations may be due to the aggregation of bio-mordant on the wool surface reducing the area for dye adsorption as some dye sites already occupied by the bio-mordant become inaccessible to dye molecules. Thus, by using chitosan as mordant for dyeing wool with cochineal, not only the ill effects of a metal mordant is eliminated, but appreciable depth of color is obtained with lower amounts of dye. Low dye absorption was observed for unmordanted wool at pH 7 which increases at pH 4 indicating acidic pH to be favorable for dyeing wool with this dye. Dye absorption for wool fiber is primarily controlled by ion-exchange reactions between the carboxyl group of dye and amino groups of wool. Below its isoelectric point (pH 4.2), wool, is positively charged, whereas above that point the carboxyl groups present in it render a net negative charge. As a result, at pH 6, the amino groups in wool will always be protonated (carboxylate anions). The pKa value for the carboxyl group of carminic acid in cochineal dye is 2.81, indicating that carminic acid will exist in carboxylate anion form at pH 4. As a result of its increased affinity, the weak carboxylate anion of dye substitutes that of the acid at pH 4. The anion of dye has a complicated character, and when it is bound on wool, it undergoes additional interactions with ionic forces, increasing wool's dyeability. However, dye absorption in wool pre-treated with chitosan followed an unanticipated pattern and showed higher dye absorption at pH 7. Generally, at pH 4, bio-mordant like chitosan acts as a cationic polyelectrolyte due to protonation of its amine groups thereby significantly increasing the dye absorption capacity of treated wool and at pH 7 it has a very low positive charge. However, the reaction between cochineal and chitosan treated wool was contrary to this indicating that the contact forces them are not solely electrostatic. Hydrogen bonding formation of carminic acid with several hydroxyl and carbonyl groups reduced in the acidic media due to protonation and loss of pair electrons of amine groups of the bio-mordant, resulting in better dye absorption in neutral medium. L* (lightness/darkness) decreased on mordanting indicating darker shades on chitosn pre-mordanted wool dyed with cochineal extract. The a* values were positive indicating redder shades, which decreases on mordanting with chitosan. The b* value of wool dyed with cochineal without any mordant was negative indicating bluer tone. These values were positive and the yellowness of the shades increased (decrease in blueness) when wool was pre-mordanted with chitosan before dyeing with cochineal extract.

5.3.3 Application of cochineal in food coloration

Carmine has a color that is similar to cured pork [82]. Cochineal-derived colors are commonly found in alcoholic beverages, yoghurts, juices, ice creams, and confectionary, but they can also be found in jams and some processed meat items [83]. Typical applications of carmine dye in food are sausages and salami displaying an intense red color [84].

5.3.3.1 Color produced

Cochineal produces intense purple color and the scarlet red color is obtained on complexing with aluminum.

5.3.3.2 Extraction & application conditions

For foodstuffs, extraction conditions for cochineal/carminic acid generally involve acid and/or enzymatic hydrolysis with or without solid-phase extraction (SPE). Carminic acid from cochineal is precipitated onto an alumina hydrate substrate. The precipitated complex called carmine is dried, grounded, and used as a food colorant. Though insoluble in water, carmine can be rendered water-soluble by reaction with a strong alkali. The color of carmine is dependent on the pH; at pH–4 and below, it is orange in color; as pH increases, it becomes redder and bluer until it becomes purplish-red above pH–6.5. The color pigment shows excellent heat and light stability.

5.3.3.3 Functional properties related to food application

Although carminic acid does not produce any genotoxic or cytotoxic effects, it has been related to cause anaphylactic reactions, asthma, urticaria, and angioedema in many individuals.

5.3.3.4 Case study-11

Surimi, minced beef, and milk were colored with naturally occurring carminic acid to change their color. Color modulation of carminic acid and carminic aluminum lake colored surimi, minced meat, and milk through the addition of different food additives, proteins, and metal ions was assessed [85]. Carminic acid rendered a light purple color to surimi while carminic aluminum lake rendered a magenta color. Minced meat and milk turned red and gray-green respectively with carminic acid. Iron and copper changed the color of the samples significantly. Changes were also observed in the case of the presence of food additives. The presence of myofibrillar protein, whey protein isolates, and soy protein isolate changed the pH of the medium resulting in a red color. Sodium nitrite is used as a preservative in the meat industry and as a chromogenic agent as well. Carminic acid changed to yellow with the addition of sodium nitrite though no change was observed in the case of the carminic aluminum lake. Also, no change in color was observed for ascorbic acid. Due to the chelation of the dye in presence of calcium ions, the color of the foodstuff changed. Hence, this dye was not found suitable for food samples rich in calcium and iron.

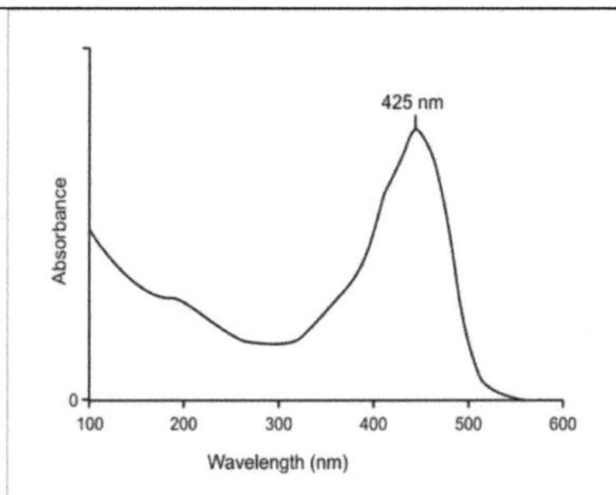

UV-Vis curve of curcumin in turmeric

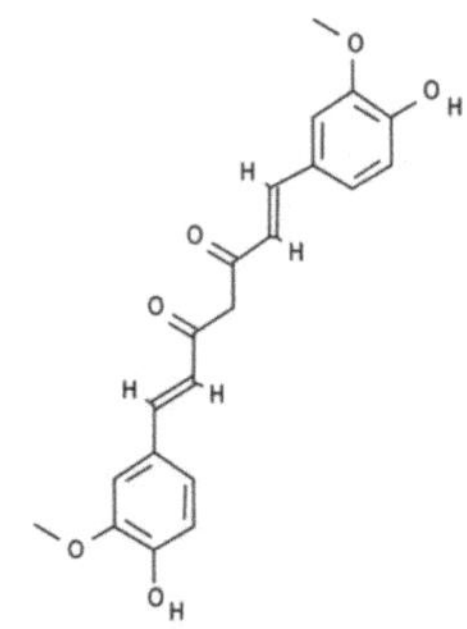

Structure of curcumin in *Curcuma longa* (turmeric)

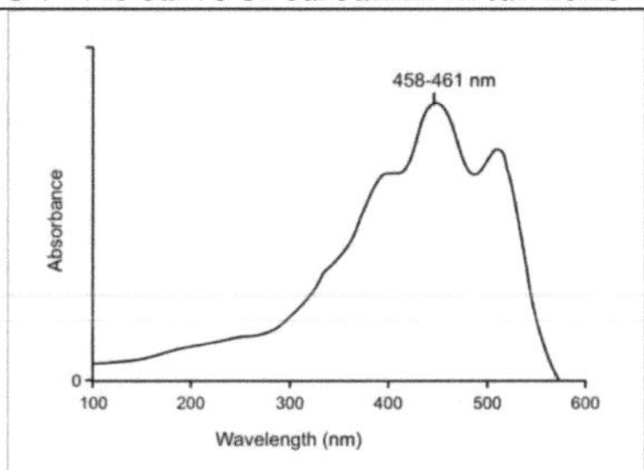

UV-Vis curve of bixin in annatto

(a)

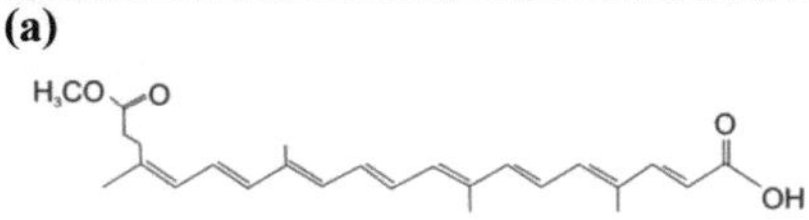

(b)

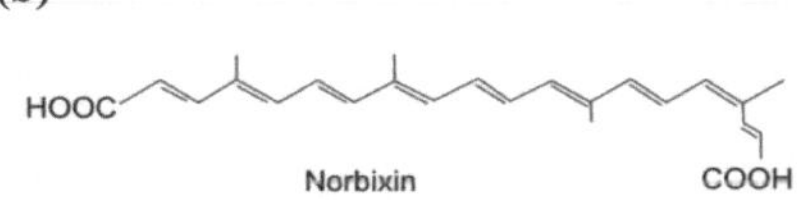

Structure of (a) bixin and (b) norbixin, in *Bixa orellana* (annatto)

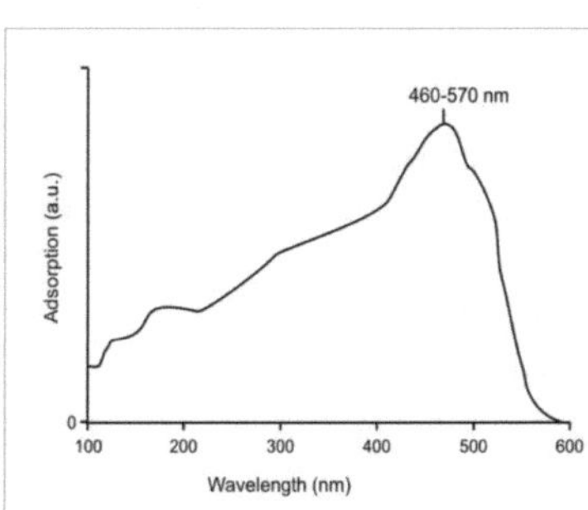

UV-Vis curve of carminic acid in cochineal

(a)

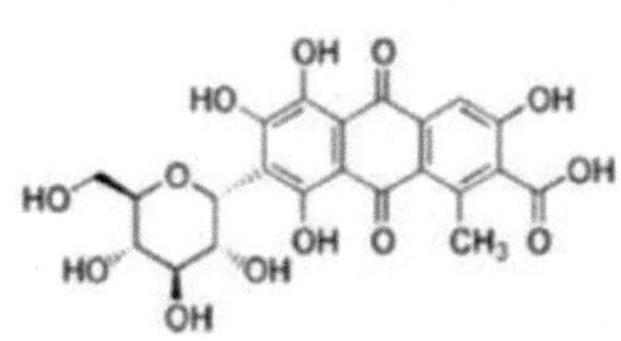

(b)

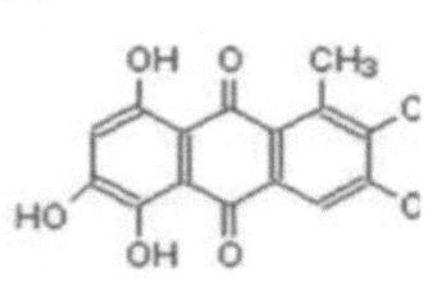

(c)

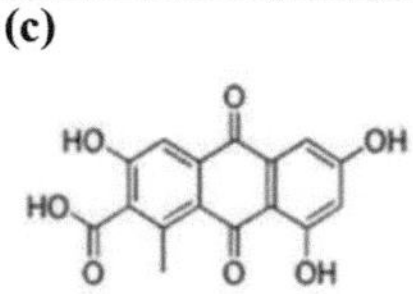

Structure of (a) carminic acid, (b) kermesic acid and (c) flavokermesic acid, in *Dactylopius coccus* (cochineal)

Figure 2.
UV–vis curves in the visible range with λ_{max} values of aqueous extracts, and chemical structures of the coloring pigments present in the source of different natural colors.

Sample	Concentration of carminic acid	
	Differential Pulse Polarography	**UV- visible spectrophotometry**
Milk (μg carminic acid /mL milk)	121 ± 4	—
Candy (mg carminic acid/g candy)	28.4 ± 1.5	27.1 ± 2.5

Table 12.
Determination of carminic acid in strawberry-flavored milk and candy using differential pulse polarography and UV–visible spectrophotometry.

5.3.3.5 Case study-12

Pulse polarography was used to quantify carmine food dye in strawberry-flavored milk and candies and the results were compared with the UV–visible spectrophotometric analysis [77]. A pH 2.0 Britton-Robinson (B-R) buffer solution was used to perform differential pulse polarography on a falling mercury electrode (peak at 489 mV). Strawberry flavored milk and candy samples were added into the polarographic cell containing B-R buffer (pH 2.0) and polarograms were taken. The concentrations were measured using the standard addition method. To compare the validity of this electroanalytical method, the samples were analyzed using UV–visible spectrophotometry (**Figure 2**). The relationship between the peak current and carminic acid concentration was linear in the range of 1 μM to 90 μM with a detection limit of 0.16 μM. The results of both methods showed similar accuracy and precision. The pulse polarographic method was advantageous as it showcased high sensitivity, low limit of determination, simple instrumentation, and easy operation (**Table 12**). The UV-vis curves with the peak of maximum absorbance of turmeric [7], annatto [41] and cochineal [75] along with chemical structures of the main coloring component present in turmeric [9], annatto [50] and cochineal [58] are given in **Figure 2**.

6. Conclusion

With the introduction of synthetic dyes like aniline, alizarin, and indigo in the mid-1800, natural dyes lost their economic and commercial significance. Synthetic dyes now dominate the market due to their wide range of colors, ease of production, and excellent fastness features. Existing limitations and technical problems in the procurement of natural dyes have further compelled the shifting of focus from natural dyes to synthetic dyes. However, within a period of 150 years, some serious drawbacks associated with synthetic dyes have come to light; synthetic dyes are suspected to release harmful chemicals that are allergic, carcinogenic, and detrimental to human health. The use of eco-friendly natural dyes that are fairly non-polluting, automatically harmonizing, more challenging, and have rare color ideas in textile and food applications is now becoming increasingly popular due to the strict environmental requirements set on the harmful chemicals used in synthetic dye production. Renewability and eco-friendliness are the two major reasons that have led to the revival of these dyes and their gradual replacement with synthetic colorants.

Author details

Deepali Singhee[1*] and Adrija Sarkar[2]

1 Departmental Textiles, Clothing and Fashion Studies, J.D. Birla Institute (affiliated with Jadavpur University), Kolkata, India

2 Department of Food Science and Nutrition Management, J.D. Birla Institute (affiliated with Jadavpur University), Kolkata, India

*Address all correspondence to: deepalisingheejdbi@gmail.com

References

[1] Choudhury AKR. Colour measurement instruments (221-269). In: Principles of Colour and Appearance Measurement. Sawston, UK: Woodhead Publishing; 2014

[2] Samanta AK, Agarwal P, Singhee D, Dutta S. Application of single and mixtures of red sandalwood and other natural dyes for dyeing of jute fabric: Studies on colour parameters/colour fastness and compatibility. Journal of the Textile Institute. 2009;**100**(7): 565-587

[3] Nanda B, Patra SK. Potentiality of natural dyes in Orissa. In: Gupta D, Gulrajani ML, editors. *Proceeding of Convention of Natural Dyes*. Delhi (India): Department of Textile Technology, IIT-Delhi; 1999. pp. 50-55

[4] Adeel S, Hussaan M, Rehman F, Habib N, Salman M, Naz S, et al. Microwave-assisted sustainable dyeing of wool fabric using cochineal-based carminic acid as natural colorant. Journal of Natural Fibers. 2018;**16**(6): 1-9

[5] Umbreen S, Ali S, Hussain T, Nawaz R. Dyeing properties of natural dyes extracted from turmeric and their comparison with reactive dyeing. Research Journal of Textile and Apparel. 2008;**12**(4):1-11

[6] Madhusankha GDMP, Thilakarathna RCN, Liyanage T, Navaratne SB. Analysis of curcumin content in Sri Lankan and Indian turmeric rhizomes and investigating its impact on the colour. International Journal of Food Science and Nutrition. 2018;**3**(4):3-05

[7] Bhardwaj N, Dadsena U. Extraction and evaluation of dyeing quality of natural curcumin. Journal of Innovations in Applied Pharmaceutical Sciences. 2017;**2**(3):1-3

[8] Sundrarajan M, Rukmani A, Gandhi RR, Vigneshwaran S. Eco friendly modification of cotton using enzyme and chitosan for enhanced dyeability of Curcuma longa. Journal of Chemical and Pharmaceutical Research. 2012;**4**(3):1654-1660

[9] Bhatti IA, Adeel S, Jamal MA, Safdar M, Abbas M. Influence of gamma radiation on the colour strength and fastness properties of fabric using turmeric (*Curcuma longa L.*) as natural dye. Radiation Physics and Chemistry. 2010;**79**(5):622-625

[10] Naveed R, Bhatti AA, Adeel S, Ashar S, Sohail I, Khan MUH, et al. Microwave-assisted extraction and dyeing of cotton fabric with mixed natural dye from pomegranate rind *(Punica Granatum L.)* and turmeric rhizome *(Curcuma longa L.)*. Journal of Natural Fibers. 2020;**3**:1-8

[11] Ghoreishian SM, Maleknia L, Mirzapour H, Norouzi M. Antibacterial properties and color fastness of silk fabric dyed with turmeric extract. Fibers and Polymers. 2013;**14**(2): 201-207

[12] Adeel S, Bhatti IA, Kausar A, Eman O. Influence of UV radiations on the extraction and dyeing of cotton fabric with *Curcuma longa L*. Indian Journal of Fibre and Textile Research. 2012;**37**(1):87-90

[13] Mirjalili M, Karimi L. Antibacterial dyeing of polyamide using turmeric as a natural dyc. Autcx Rcscarch Journal. 2013;**13**(2):51-56

[14] Sachan K, Kapoor VP. Optimization of extraction and dyeing conditions for traditional turmeric dye. Indian Journal of Traditional Knowledge. 2007;**6**(2): 270-278

[15] Samanta AK, Singhee D, Sethia M. Application of single and mixture of selected natural dyes on cotton fabric: A scientific approach. Colourage. 2003; **50**(10):29-42

[16] Worku A. Extraction and application of natural dye on tanned leather and eco-friendly approach. International Research Journal of Engineering and Technology. 2018;**5**(8): 431-434

[17] Gupta D, Jain A, Panwar S. Anti-UV and anti-microbial properties of some natural dyes on cotton. Indian Journal of Fibre and Textile Research. 2005;**30**(2): 190-195

[18] Kunkely H, Vogler A. Absorption and luminescence spectra of cochineal. Inorganic Chemistry Communications. 2011;**14**(7):1153-1155

[19] Sriumaoum V, Sodsangchan C, Setthayanond J, Suwanruji P, Tooptompong P. Effect of chitosan and turmeric dye on ultraviolet protection properties of polyester fabric. In: Applied Mechanics and Materials. Vol. 535. Country: Trans Tech Publications Ltd.; 2014. pp. 658-661

[20] Hosen MD, Rabbi MF, Raihan MA, Mamum MAA. Effect of turmeric dye and biomordants on knitted cotton fabric coloration: A promising alternative to metallic mordanting. Cleaner Engineering and Technology. 2012;**3**:100124

[21] Delfanian M, Sahari MA. Improving functionality, bioavailability, nutraceutical and sensory attributes of fortified foods using phenolics-loaded nanocarriers as natural ingredients. Food Research International. 2020;**137**: 109555-109577

[22] Lodh J, Khamrui K, Prasad WG. Optimization of heat treatment and curcumin level for the preparation of anti-oxidant rich ghee from fermented buffalo cream by central composite rotatable design. Journal of Food Science and Technology. 2018;**55**(5):1832-1839

[23] Al-Obaidi LFH. Effect of adding different concentrations of turmeric powder on the chemical composition, oxidative stability and microbiology of the soft cheese. Plant Archives. 2019; **19**(2):317-321

[24] Zanzer YC, Batista ÂG, Dougkas A, Tovar J, Granfeldt Y, Östman E. Difficulties in translating appetite sensations effect of turmeric-based beverage when given prior to isoenergetic medium-or high-fat meals in healthy subjects. Nutrients. 2019; **11**(4):736-742

[25] Yildiz E, Gungor G, Yilmaz H, Gocmen D. Changes in bioaccessibility, phenolic content and antioxidant capacity of novel crackers with turmeric (*Curcuma longa L.*) and mahaleb (*Prunus mahaleb* L.) powders. Quality Assurance & Safety of Crops and Food. 2019;**11**(2):107-116

[26] Janiszewska-Turak E, Pisarska A, Królczyk JB. Natural food pigments application in food products. Nauka Przyroda Technologie. 2016;**10**(4):1-13

[27] Verma K, Manorama, Pophaly SD. Natural food colours. Plant Archives. 2018;**18**(1):1159-1162

[28] Mohamed MF, Dailin DJ, Gomaa SE, Nurjaydi GM, Enhasy HE. Natural colorant for food: A healthy alternative. International Journal of Scientific and Technology Research. 2019;**8**(11):3161-3166

[29] Prathapan A, Lukhman M, Chami A, Sundaresan A, Raghu KG. Effect of heat treatment on curcuminoid, colour value and total polyphenols of fresh turmeric rhizome. International Journal of Food Science and Technology. 2009;**44**(7):1438-1444

[30] Jung YN, Hong J. Changes in chemical properties and bioactivities of turmeric pigments by photo-degradation. AIMS Agriculture and Food. 2021;**6**(2):754-767

[31] Vilar DDA, Vilar MSDA, Raffin FN, Oliveira MRD, Franco CFDO, de Athayde-Filho PF, et al. Traditional uses, chemical constituents, and biological activities of *Bixa orellana L.*: A review. The Scientific World Journal. 2014;**2014**:857292:1-11

[32] Islam OE. Greener natural dyeing pathway using a by-product of olive oil; prina and biomordants. Fibers and Polymers. 2017;**18**(4):773-785

[33] Ntohogian S, Gavriliadou V, Christodoulou E, Nanaki S, Lykidou S, Naidis P, et al. Chitosan nanoparticles with encapsulated natural and uf-purified annatto and saffron for the preparation of UV protective cosmetic emulsions. Molecules. 2018;**23**(9):2107

[34] Chavan RB, Singh TG, Kaur B. Dyeing of cotton with Indian and Brazil annatto. In: Gupta D, Gulrajani ML, editors. Proceeding of Convention of Natural Dyes. Department of Textile Technology: IIT Delhi; 2001. pp. 58-63

[35] Sarkar AK, Seal CM. Colour strength and colour fastness of flax fabrics dyed with natural colourants. Clothing and Textiles Research Journal. 2003;**21**(4): 162-166

[36] Devi A, Katyayini VKLT, Sumanthy BS. Annato - A bright natural colourant for cotton. Textile Trends. 2002;**45**(1):29-34

[37] Javali UC, Sreenivasa, Radhalakshmi YC. Application of natural dye - Annato on mulberry silk. Colourage. 2009;**56**:50-52

[38] Gulrajani M, Bhaumik S, Oppermann W, Hardtmann G. Kinetic and thermodynamic studies on red sandalwood. Indian Journal of Fibre and Textile Research. 2002;**27**(1):91-94

[39] Selvi T, Rathinam A, Madhan B, Raghava RJ. Studies on the application of natural dye extract from Bixa orellana seeds for dyeing and finishing of leather. Industrial Crops and Products. 2013; **43**(1):84-86

[40] Devi M, Ariharan VN, Prasad NP. Annatto: An ecofriendly and potential source for natural dye. International Research Journal of Pharmacy. 2013; **4**(6):106-108

[41] Das D, Maulik SR, Bhattacharya SC. Dyeing of wool and silk with bixa orellana. Indian Journal of Fibre and Textile Research. 2007;**32**(3):366-372

[42] Rincón CTS, Montoya JEZ, Gómez GLC. Optimizing the extraction of phenolic compounds from Bixa orellana L. and effect of physicochemical conditions on its antioxidant activity. Journal of Medicinal Plant Research. 2014;**8**(46):1333-1339

[43] Preston HD, Rickard MD. Extraction and chemistry of annatto. Food Chemistry. 1980;**5**(1):47-56

[44] Erdawati E, Allanas A, Wesnina W. Extraction of bixin from annatto seeds with microwave. In: Journal of Physics: Conference Series, Vol-1869, 2nd Annual Conference of Science and Technology. Malang: Indonesia; 2020

[45] Kawlekar SR, Mukundar U. Dyeing of cotton fabric with annatto (Bixa orellana). International Dyer. 2011; **196**(10):37

[46] Yolmeh M, Najafi MBH, Farhoosh R, Salehi F. Modeling of antibacterial activity of annatto dye on escherichia coli in mayonnaise. Food Bioscience. 2014;**8**:8-13

[47] Islam S, Rather LJ, Mohammad F. Phytochemistry, biological activities and

potential of annatto in natural colorant production for industrial applications – A review. Journal of Advanced Research. 2016;**7**(3): 499-514

[48] Yolmeh M, Habibi-Najafi MB, Shakouri S, Hosseini F. Comparing antibacterial and antioxidant activity of annatto dye extracted by conventional and ultrasound-assisted methods. Zahedan Journal of Research in Medical Sciences. 2015;**17**(7):29-33

[49] Pagnan CS, Mottin AC, Oréfice RL, Ayres E, Câmara JJD. Annatto-colored poly (3-hydroxybutyrate): A comprehensive study on photodegradation. Journal of Polymers and the Environment. 2018;**26**(3): 1169-1178

[50] Dumitrescu I, Vankar PS, Srivastava J, Mocioiu AM, Iordache OG. Dyeing of cotton, silk and wool with Bixa orellana in the presence of enzymes. Industrial Texting. 2012; **63**(6):1-7

[51] Silva LJG, Pereira ARS, Pereira AMPT, Pena A, Lino CM. Carmines (E120) in colored yoghurts: A case-study contribution for human risk assessment. Food Additives & Contaminants: Part A. 2021;**38**(8): 1316-1323

[52] Shams NA, Dehnavi E, Hajipour A, Ekrami E. Dyeing of polyamide fibre with cochineal natural dye. Pigment & Resin Technology. 2016;**45**(4):252-258

[53] Haila KM, Lievonen SM, Heinonen MI. Effects of lutein, lycopene, annatto, and γ-tocopherol on autoxidation of triglycerides. Journal of Agricultural and Food Chemistry. 1996; **44**(8):2096-2100

[54] Balaswamy K, Pamidighantam P, Babu P, Akula S. Application of annatto (*Bixaorellena L.*) dye formulations in Indian traditional sweetmeats: Jilebi and jangri. Indian Journal of Traditional Knowledge. 2012;**11**(1):103-108

[55] Mala KS, Rao PP, Prabhavathy MB, Satyanarayana A. Studies on application of annatto (Bixa orellena L.) dye formulations in dairy products. Journal of Food Science and Technology. 2015; **52**(2):912-919

[56] Deveoglu O. A review on cochineal (*Dactylopius Coccus Costa*) dye. Research Journal of Recent Sciences. 2020;**9**(3): 37-43

[57] Francis FJ. Less common natural colorants (311-341). In: Hendry GAF, Houghton JD, editors. Natural Food Colorants. 2nd ed. Bishopbriggs, Glasgow: Blackie Academic & Professional-Chapman & Hall; 1996

[58] Gonzlez M, M'ndez J, Carnero A, Lobo MG, Alfonso A. Optimizing conditions for the extraction of pigments in cochineals (*Dactylopius coccus Costa*) using response surface methodology. Journal of Agricultural and Food Chemistry. 2002;**50**(24): 6968-6974

[59] Stintzing FC, Carle R. Cactus stems Opuntia spp.: A review on their chemistry, technology, and uses. Molecular Nutrition & Food Research. 2005;**49**:175-194

[60] Kendrick A. Natural food and beverage colourings. In: Natural Food Additives, Ingredients and Flavourings. 2021

[61] Mendez J, Gonzalez M, Lobo MG, Carnero A. Color quality of pigments in cochineals (*Dactylopius coccus* Costa). Geographical origin characterization using multivariate statistical analysis. Journal of agricultural. Food Chemistry. 2004;**52**(5):1331-1337

[62] Ke G, Zhu K, Chowdhury MH. Dyeing of cochineal natural dye on cotton fabrics treated with oxidant and

chitosan. Journal of Natural Fibers. 2021;**18**(3):317-329

[63] Arroyo-Figueroa G, Ruiz-Aguiar GML, Cuevas-Rodriguez G, Gonzalex G. Cotton fabric dyeing with cochineal extract: Influence of mordant concentration. Coloration Technology. 2010;**127**(1):39-46

[64] Amin N, Rehman F, Adeel S, Ahmad T, Muneer M, Haji A. Sustainable application of cochineal-based anthraquinone dye for the coloration of bio-mordanted silk fabric. Environmental Science and Pollution Research. 2020;**27**(7): 6851-6860

[65] Gawish M, Reham F, Ramadan M, Mashaly H. Eco-friendly multifunctional properties of cochineal and weld for simultaneous dyeing and finishing of proteinic fabrics. International Journal of Engineering and Technology. 2016;**8**(5):2246-2253

[66] Shahid-ul-Islam LJ, Mohammad F. Phytochemistry, biological activities and potential of annatto in natural colorant production for industrial applications–A review. Journal of Advanced Research. 2016;7(3):499-514

[67] Marczenko Z, Balcerzak M. Separation, Preconcentration and spectrophotometry in inorganic analysis (chapter 11, 121-128). In: Kloczko E, editor. Analytical Spectroscopy Library. Vol. 10. Elsevier Science; 2000

[68] Luo Y, Li M, Du J. Esterification of cochineal carmine used for dyeing cationic modified cotton with high color fastness. Journal of Engineered Fibers and Fabrics. 2017;**12**(2):60-65

[69] Valipour P, Ekrami E, Shams-Nateri A. Colorimetric properties of wool dyed with cochineal: Effect of dye-bath pH. Progress in Color, Colorants and Coatings. 2014;**7**:129-138

[70] Lee DM, Jeon DW, Jong J. Effect of chitosan treatment methods on the dyeing of cotton, nylon, and PET using cochineal (II) - focusing on color change by laundering, wash fastness and abrasion fastness. Fashion Business. 2005b;**9**(2):71-83

[71] Bae JS, Huh MW. The dye ability and antibacterial activity of wool fabric dyed with cochineal. Textile Coloration and Finishing. 2006b;**18**(5):22-29

[72] Kamel MM, El-Zawahry MM, Ahmed NSE, Abdelghaffar F. Ultrasonic dyeing of cationized cotton fabric with natural dye. Part 1: Cationization of cotton using Solfix E. Ultrasonics Sonochemistry. 2009;**16**:243-249

[73] Lee DM, Jeon DW, Jong J. Effect of chitosan treatment methods on the dyeing of cotton, nylon, and PET using cochineal (III) - light fastness and perspiration fastness characteristics. Fashion Business. 2005a;**9**(3):99-113

[74] Velíšek J, Davídek J, Cejpek K. Biosynthesis of food constituents: Natural pigments. Part 1—A review. Czech Journal of Food Sciences. 2007; **25**(6):291-315

[75] Ammayappan L, Shakyawar DBB. Dyeing of carpet woolen yarn using natural dye from cochineal. Journal of Natural Fibers. 2016;**13**(1):42-53

[76] Bae JS, Huh MW. Natural dyeing properties and antibacterial activity of nylon fabric dyed with cochineal. Fashion & Textile Research Journal. 2006a;**8**(6):702-708

[77] Bae JS, Kim YK, Huh MW. The dye ability and antibacterial activity of silk, rayon fabrics dyed with cochineal. Textile Coloration and Finishing. 2006; **18**(6):1-9

[78] Trotman ER. Dyeing and Chemical Technology of Textile Fibers (187–217). 6th ed. London: Edward Arnold; 1984

[79] Kumar ANA, Joshi G, Ram HYM. Sandalwood: History, uses, present status and the future. Current Science. 2012;**103**(12):1408-1416

[80] Velmurugan P, Selvi AT, Lakshmanaperumalsamy P, Park J, Oh BT. The use of cochineal and *Monascus purpureus* as dyes for cotton fabric. Coloration Technology. 2013; **129**(4):246-251

[81] Mehrparvar L, Safapour S, Sadeghi-Kiakhani M, Gharanjig K. Chitosan-polypropylene imine dendrimer hybrid: A new ecological biomordant for cochineal dyeing of wool. Environmental Chemistry Letters. 2016; **14**(4):533-539

[82] Sesselman U, Brunner E, Zellner M. Farbstoffe in Fleischerzeugnissen-Neue Moglichkeiten aufgrund der EG-Richtlinie uber Farbstoffe, die in Lebensmitteln verwendet warden durfen. Fleischwirtschaft. 1995;**75**(11): 1288-1291

[83] Villano D, Garcıa-Viguera C, Mena P. Colors: Health effects. Encyclopedia of Food and Health. 2016; **2**:265-272

[84] Ruiz-Capillas C, Tahmouzi S, Triki M, Rodriguez L, Kimenez-Colmenero F, Herrero AM. Nitrite-free Asian hot dog sausages reformulated with nitrite replacers. Journal of Food Science and Technology. 2015;**52**(7): 4333-4341

[85] Liu Q, He Z, Zeng M, Qin F, Wang Z, Liu G, et al. Effects of different food ingredients on the color and absorption spectrum of carminic acid and carminic aluminum lake. Food Science & Nutrition. 2021;**9**:36-43

Section 4

Colour Measurement and Applications for Paint and Pharmaceutical Industry

Chapter 7

A Digital Image-Based Colorimetric Technique Use for Quantification of Green Active Pharmaceuticals Obtained from Natural Sources

Vitthal V. Chopade and Jayashri V. Chopade

Abstract

Colorimetry is the determination of colors, as name indicates. This method can use for to find out the concentration of compound (solute) in a colored solution in terms of chemical analysis (solvent). We frequently need to quantify the quantity of a specific component in a combination or the concentration of a solution during scientific activity. The trick is to determine the color differences between various combinations and their absolute values. This is more instructive and scientifically valuable than relying on subjective judgments like the color of the solution. Colorimetry is the measurement of colors, as the name implies. It is the measurement of the concentration of a certain compound (solute) in a colored solution in terms of chemical analysis (solvent). We frequently need to quantify the quantity of a specific component in a combination or the concentration of a solution during scientific activity. The trick is to determine the color differences between various combinations and their absolute values. This is more instructive and scientifically valuable than relying on subjective judgments like the color of the solution. Colorimetry is used in a digital image-based (DIB) approach for determining active medicinal components. A computerized scanner with a controlled light intensity was connected to the detector. Different histograms were used to transform the photos. The colorimetric analysis of digital images provided for an easy-to-use and ecologically friendly method.

Keywords: colorimeter, digital image based (DIB) colorimetric analysis, quantification of color, UV-vis spectrophotometer, reflectance spectrophotometer, green active pharmaceuticals

1. Introduction

When a light of occurrence with intensity (Io) passes from a solution, a portion of the light is revealed (Ir), a portion is absorbed (Ia), and the rest is transmitted (It),

Thus,

$$\mathrm{Ir} + \mathrm{Ia} + \mathrm{It} = \mathrm{Io}$$

Because of the measurement of (Io), (Ir) is omitted in colorimeters, and it is sufficient to regulate the (Ia). The light will be replicated (Ir) is kept consistent for this purpose by using cells with identical characteristics. The values of (Io) and (It) are then calculated.

The two fundamental equations of photometry, on which the colorimeter is based, show the mathematical link between the amount of light absorbed and the concentration of the substance.

Beer's Law is a rule that states that if you drink.

The amount of light absorbed is directly proportional to the concentration of the solute in the solution, according to this rule.

$$I0/It = asc$$
$$Log10\ (I0/It) = asc$$

where,
as = Absorbency index.
C = Solution Concentration.

2. Lambert's rule

According to Lambert's law, the amount of light absorbed is proportional to the length and thickness of the solution under investigation [1–5].

$$I0/It = asb$$
$$A = \log 10\ (I0/It) = asb$$

where,
A = Test Absorbance.
as = Standard absorbance.
b = the solution's length/thickness.

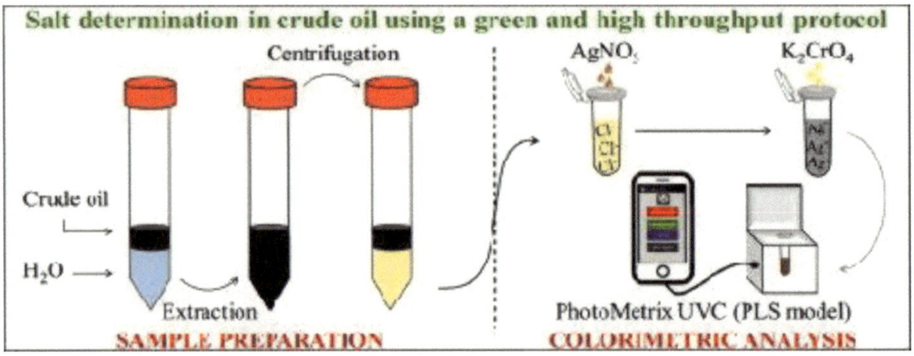

3. Instrumentation

Components for Colorimeter,

The Colorimeter is a technique that measures the amount of absorbed light of a given wavelength in order to determine the concentration of a sample. Duboscq Colorimeter, created by Jules Doboscq in 1870, is one of the first and most popular designs. The components of a colorimeter are as follows:

1. A source of light to illuminate the solution, commonly a blue, green, or red LED.

2. Filters for light wavelengths of red, blue, and green.

3. A slit through which a light beam can be focused.

4. A condenser lens that divides a light beam into parallel rays.

5. A cuvette in which the solution will be held. Glass, quartz, or plastic are used to construct.

6. A photoelectric cell, which is a vacuum-filled cell that measures and converts transmitted light into electrical output. Light-sensitive materials, such as selenium, are used to create these.

7. A transmittance or absorbance meter, either analogue (e.g., a galvanometer) or digital (e.g., a multimeter) [6].

4. Colorimeter's process

1. The LED or tungsten bulb emits white light to begin the procedure.

2. A condenser lens concentrates this light into a parallel beam after passing through a slit.

3. A series of lenses focus a certain wavelength of light on the solution as the beam passes through them.

4. The problem answer is kept in a cuvette with a predetermined path length (width of the cuvette). The solution absorbs a portion of the light and transmits the remainder.

5. The intensity of the transmitted light is transformed into an electrical signal, which is recorded by an analogue (e.g., a galvanometer) or digital meter [7].

5. The colorimetric principle

Colorimetry is the measuring of electromagnetic radiation's wavelength and intensity in the visible spectrum. It's widely used for determining the quantities of light-absorbing compounds and identifying them. The equation log can be used to represent the two laws when they are combined. I0/I = kcd, where I0 is the incident light intensity, I is the transmitted light intensity, c is the absorbing substance concentration, d is the distance through the absorbing solution, and k is a constant that varies depending on the absorbing substance, the wavelength of light used, and the units used to specify c and d.

The intensity of radiation passed through layers of various thicknesses of two solutions of the same absorbing substance, one with a known concentration and the other unknown, is a straightforward application of this term. The equipment is known as a photoelectric colorimeter when it compares intensities using a photoelectric cell rather than the eye.

The comprehensive noticeable band (white light) is frequently employed in colorimetry, The flattering color of the one captivated is seen as transmitted light. The instrument is called a spectrophotometer if it uses homochromatic light or a constrained band of radioactivity. It is used to demeanor dimensions in the ultraviolet and infrared regions as well as the visible spectrum. Colorimeters have mostly been superseded by spectrophotometers [8].

6. Color grading

Colorimetry is a term for the measuring of color. This field employs a wide range of instruments. The most advanced, spectrophotometers, measure the quantity of energy in each spectral wavelength. The emittance curves for light sources (see figure) and the reflectance curve for the paint color emerald green (see figure) are typical spectrophotometer results.

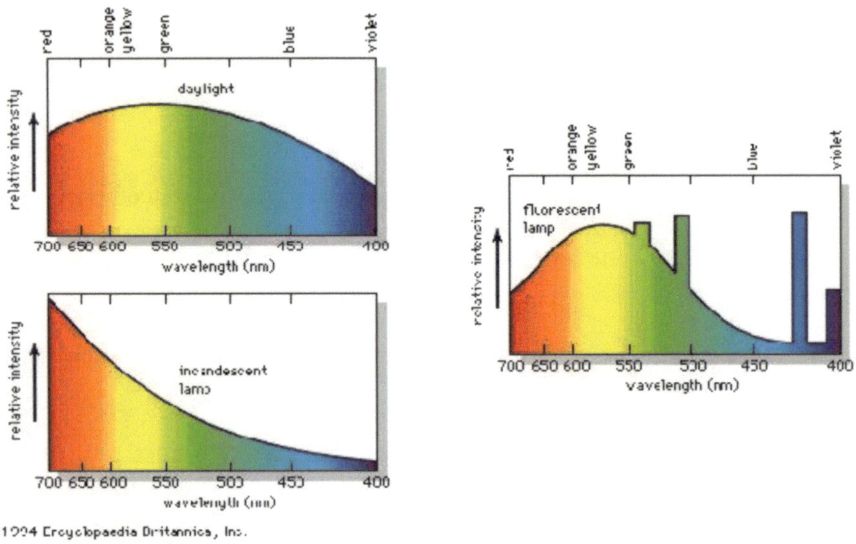

Electromagnetic radiation is defined as the movement of photons (also known as light quanta) over space, according to contemporary quantum theory. Photons are energy packets that travel at the universal speed of light at all times. The symbol h stands for Planck's constant, and the value of corresponds to the frequency of an electromagnetic wave in classical theory. The numerical density of photons with the same energy h corresponds to the intensity of the radiation. The frequency of electromagnetic radiation influences these occurrences, as well as how it is made and viewed, how it occurs in nature, and how it is used in technology. The frequency spectrum of electromagnetic radioactivity varieties from extremely low values in the radio, television, and microwave ranges to noticeable light and elsewhere, too much sophisticated standards in ultraviolet light, X-rays, and gamma rays [9].

7. Color measurement aspects

A test bulb glows proceeding a bleached monitor, which is observed by an spectator, in a very ancient experiment. One or more of three lamps, each capable of producing light in three different colors, such as red, green, and blue, illuminate a neighboring section of the screen. The principal light is chosen at random, although it is tightly regulated. The observer can match the collective shade on the monitor to that of the test bulb by altering the intensities of these lights. The tristimulus value of a color is defined as the sum of the three primaries [10].

7.1 Color measurement is divided into three categories

i. Visual matching and spectral matching are the two types of matching. Measurement in three colors. The information representative distinctive wavelengths collected from the noticeable bands of these things allows objects to be discriminated based on information representative distinguishing wavelengths achieved from the visible spectra of these substances, rather than tristimulus values. Spectrophotometer by immovable wavelength frequencies or specific shade devices are examples of the instruments used for this purpose.

ii. There are two types of colorimetric analysis: visual colorimetry and photoelectric colorimetry. There are two distinct methods to each method. The color absorption capacity is used to determine the amount of a colored component in a solution.

iii. To determine the true shade of the material in relations of tristimulus standards, which provides the shade a humanoid-like look.

8. Color matching

a. **Visual match:** The simplest and oldest method of color measuring is visual color matching, which allows a shade to be projected but not restrained. It entails choosing or creating a shade that corresponds to the specimen's X, Y, and Z tristimulus standards. The most straightforward method is to employ a large, regularly organized collection of color samples that can be visually matched by an observer alone or with the assistance of visual colorimeters, with the observer altering the reference color until it is judged to be equivalent to the specimen. This technique must be done in a well-lit environment. Standard color matching boxes with various illuminants can be found on the market and used to perform visual matching.

b. **Spectral matching:** Spectral matching is the process of matching colors based on their wavelength The most precise approach for color measuring is spectral matching, which has two steps. The first, referred to as spectral photometry, entails measuring physical characteristics such as reflectance or transmittance curves, while the second, referred to as tristimulus assessment, entails recalculating or modifying the physical values derived from tristimulus data. Modern spectral measuring tools (spectrophotometers, spectrocolorimeters) can usually cover both stages in one run. By using this method, it is expected that 10 million hues may be distinguished. A monochromator, which is typically achieved in traditional structures using a grating, is at the heart of spectral color measurement devices. The slit width influences the wavelength bandwidth. Though the effect is the same, the most common method of covering the spectrum is to rotate the grating or prism rather than move the slit. The light sensitivity of the linear array detector, as well as optical fibers and concave gratings capable of focalizing light on the focusing plane where the array detector is situated, are used in today's technology. This allows for the measurement of the entire light spectrum in a single shot while avoiding the use of spinning parts, and at a lower cost. Optical spectrophotometric scanners can now be used to do spectral matching under huge area samples. They're spectrum imaging tools that can also be used to assess sample color consistency. The spectrum reflectance of the varnished layer is measured using spectral imaging, which is a physical quantity that is independent of the measurement apparatus. An ordered collection of the spectral reflectance of its pixels can be stored as a digital image based on this physical quantity. Spectral pictures have a number of benefits, including the following:

1. Image reconstruction in CIE color space with illuminant selection and color matching functions

2. Image replication with high fidelity using output devices such as monitors or printers that can work in more than trichromatic modes.

Colorimetry by spectrophotometer: The photoelectric method's main premise involves converting the glowing energy conveyed by a stained system to rechargeable energy, which can be detected more easily, using a photoelectric detector. A light source with a constant intensity is required by the instrument.

9. Photometry of filters

a. Color filters are used as a wavelength selection in this method, as indicated by the name. This strategy relies heavily on the selection of an appropriate filter. To increase sensitivity, the filter's color should complement the color of the solution being tested. Most filter photometers produce a spectral bandwidth of 50 mj.i (or nm). (A micrometer is an insufficiently small unit.) Despite the fact that filter photometry provides more accurate and exact findings than the visual approach, it only covers a small portion of the spectrum compared to the spectrophotometer. As a result, shortened spectrophotometer is a term used to describe them.

b. Spectrophotometry: A spectrophotometer measures a sample's reflectance or transmittance as a function of wavelength. Which is a monochromatic that creates a spatially dispersed colored light spectrum on its focal plane. This allows you to choose any wavelength of incident radiant energy and determine the spectrophotometric curves of color samples radially. Interference from other color compounds can frequently be avoided by choosing the right wavelength.

c. Trichromatic Colorimetry: According to researchers, the human eye has three receptors, and the differences in their responses contribute to color perceptions. Grassmann and Maxwell, on the other hand, were the first to assert unequivocally that color can be described mathematically in terms of three independent variables.

Color values and color difference measurements can be expressed in the following ways:

Newer benchtop instruments are also smaller, more automated, and less expensive than their forerunners. A new generation of entry-level benchtops has prices that are equivalent to portable computers. Software tools for color formulation and quality control have also made significant progress. Windows is used in the most, if not all, of the new debuts. They enable for customized display screen configurations and present electronically generated simulations of color standards and sample formulations in a variety of lighting conditions. Spreadsheet functionality, e-mail data transfer, and data transfer to and from portable spectrophotometers are among the other new features that make data manipulation easier [11].

10. Conclusion

Colorimetry has long been used in experimental, décor, pharmaceuticals, cosmetics, textiles, surface coatings, paper, ink, plastic, color photography, glass, and paint, as well as in biology as a health indicator and for cell and tissue examination. Color measurement is complicated by the fact that one can only 'measure' the reflection and transmission qualities of the materials that produce the color.

There are two methods that could be used: (i) color capacity of an purpose; (ii) strength quantity of reflected/communicated light at specific wavelengths The goal of this research was to create a tristimulus colorimeter with red, green, and blue leds, as well as to investigate their various uses. The architecture of the human visual system, color vision, and color vision theories are all covered in length in this chapter. We begin with an overview of color science and the internationally recommended CIE system, which serves as the foundation for colorimetric application, before moving on to color measurement instruments and color spaces used in color calculation and correction. This technique is used to quantify green active pharmaceuticals derived from natural sources.

Author details

Vitthal V. Chopade[1]* and Jayashri V. Chopade[2]

1 Progressive Education Society's, Modern College of Pharmacy, Pune, India

2 Pimpri Chinchwad College of Engineering and Research, Pune, India

*Address all correspondence to: vitthalchopade@gmail.com

References

[1] Othmer K. "Encyclopedia of Chemical Technology", Vol. 5, 2nd edn, Interscience Publishers, 1964. Color and Construction of Organic Dyes, 763-765, Color measurement 809-811

[2] Henderson ST, Marsden. Lamps and Lighting, A Manual of Lamps and Lighting. Edward Arnold Publishers; 1975. pp. 45-87. Ch 3 and 4

[3] Heinz-Helmut Perkampus, "Encyclopedia of Spectroscopy", 1993, ch1, pp. 88-101 and 514-515

[4] Hunt RWG. Measuring Color. Chichester, UK: Ellis Horwood; 1987

[5] MacAdam DL. Color Measurement. Springer, Berlin: Theme and Variations; 1981

[6] Hanbury A. The Taming of the Hue, Saturation and Brightness Color Space. France: Centre de Morphologie; 1998. pp. 234-243

[7] Wyszecki G, Stiles WS. Color Science, Concepts and Methods, Quantitative Data and Formulas. New York: Wiley; 1967

[8] Oleari C. Standard Colorimetry: Definitions, Algorithms, and Software. West Sussex, England: Wiley; 2016

[9] LutfiFirdausa M, WiwitAlwib, FerliTrinoveldib, ImanRahayuc, Rahmidard L, KanconoWarsitoa. Determination of Chromium and Iron Using Digital Image-based Colorimetry. Procedia Environmental Sciences. 2014; **20**:298-304

[10] Savitribai Phule Pune University Pune. Literature data Chapter Two/Literature review on colorimetic system. In: Chemistry Literature Book. Pune: Pune University; 2015. pp. 11-63

[11] Dakashev AD, Pavlov SV, Stancheva KA. Method of vis spectrometry based on measuring solution color, using digital camera and digital image processing. ACAIJ (Analytical Chemistry-An Indian Journal). 2013;**13**(8):303-308

Section 5

Colorimetric Evaluation and Applications in Nuclear Fusion and Camouflaged Defence Textiles

Chapter 8

Colorimetry in Nuclear Fusion Research

Gen Motojima

Abstract

Colorimetry is a unique technique among research fields. The technique is also utilized in nuclear fusion research. The motivation is to evaluate the wide range of distribution of the deposition layer on the surface of the vacuum vessel. The deposition layer affects the control of fuel particles. Therefore, the result from colorimetry can contribute to the study of particle control in fusion plasma. In a particle control study, global particle balance analysis is usually conducted. Also, long-term samples irradiated by plasma have been analyzed. Colorimetry has the role of a bridge between these analyses. In this chapter, a demonstration of colorimetry in fusion devices is introduced.

Keywords: nuclear fusion, colorimetry, color analyzer, reflection rate, deposition layer, wall retention, Large Helical Device, Wendelstein 7-X

1. Introduction

We are now facing challenging times in the global environment. While people's social lives are becoming more affluent, global changes (a global crisis, if you will) such as global warming must be solved on a global scale. The world is aiming to become carbon neutral by 2050. Here, carbon neutrality, which is often used in conjunction with the term "a decarbonised society", is a concept that aims to limit carbon dioxide emissions, which are the main cause of global warming. To achieve carbon neutrality, hydrogen is attracting attention as the energy of the new age. Hydrogen is considered to be the most abundant material in the universe and can be used as heat energy through combustion. The main advantage of the use of hydrogen is that hydrogen does not produce carbon dioxide when it is used as energy. The energy efficiency of hydrogen is so high that it is even used as fuel for rockets. Nuclear fusion energy is the key energy source, using hydrogen as a fuel.

Nuclear fusion is a reaction in which two or more nuclei are transformed into nuclei of higher atomic numbers. The easiest fusion reaction to initiate on the earth is a nuclear reaction in which helium (He) and neutrons (n) are produced from deuterium (D) and tritium (T). 17.6 MeV (=2.82×10^{-12} J) of energy is released in a single nuclear reaction. The reaction equation can be expressed as an Eq. (1).

$$ {}^{2}_{1}\mathrm{D} + {}^{3}_{1}\mathrm{T} \rightarrow {}^{4}_{2}\mathrm{He}\left(3.52\,\mathrm{MeV}\right) + {}^{1}_{0}\mathrm{n}\left(14.06\,\mathrm{MeV}\right) \tag{1} $$

The nuclear reaction of 1 g of D-T fuel corresponds to the thermal energy of the combustion of about 8 tons of fossil fuel oil. For a reaction as shown in Eq. (1) to

IntechOpen

occur, it is necessary to increase the relative velocity of the nuclei so that the kinetic energy is greater than the electric potential energy at the position where the nuclear force is effective. The relative velocity of hydrogen can be increased by heating it to a high temperature and by the increase of the thermal motion of the nuclei. For a fusion reaction to occur on earth, the hydrogen fuel particles would have to be heated to several hundred million degrees. In such a high-temperature state, the constraint of electrical attraction between nuclei and electrons is broken and they become discrete. This state is called "plasma". There are several ways to confine the plasma, but here we introduce research on fusion plasmas with the "magnetic confinement method". Few readers may be able to imagine the relevance of colorimetry to fusion research. The author hopes that this chapter will make the reader aware that colorimetry is a method that can make a significant contribution to fusion research. Especially, colorimetry helps the understanding of particle control in nuclear fusion research.

2. Contribution of colorimetry to nuclear fusion research

To gain stable fusion energy, a stable fuel supply control is necessary. For performing fuel particle control, understanding of the particle absorption process at the plasma-facing wall is an important issue, as well as the establishment of a fuel supply method. For example, in the Large Helical Device (LHD) at the National Institute for Fusion Science (NIFS) in Japan, which is one of the largest superconducting machines among helical plasma experimental devices [1], a global particle balance analysis was carried out in a 48-minute-long helium discharge with a total heating power of 1.2 MW using ion cyclotron heating and electron cyclotron heating, and dynamic wall retention of fuel particles was found [2]. It has been found that the dynamic wall retention can be explained by the temperature dependence of the particle retention of the plasma-facing walls composed of the divertor plate (carbon) and the first wall (stainless steel) [3–5]. In this discharge, 60% of the fuel particles were absorbed by the wall, and long-term samples irradiated by plasmas showed that the absorbed amount increased linearly with the thickness of the carbon-based deposition layer [5]. To understand the particle retention in the wall, it is important to identify "where" and "how much" the deposition layer is distributed in the vacuum vessel over a wide area. However, it is not practical to evaluate the thickness of the deposition layer on all the plasma-facing walls of the vacuum vessel using irradiated samples, because it takes time to analyze the samples and only a limited number of samples can be installed in the vacuum vessel. Therefore, a new method has been devised: color analysis. So far, color analysis has been carried out on plasma experimental devices such as the TEXTOR-94 and ASDEX-U [6, 7]. In these machines, the hue of the color was measured to assess the thickness of the deposition layer. For this purpose, a CCD camera was used in the TEXTOR-94 and an imaging camera was used in the ASDEX-U. Although, there have been reports of using cameras to evaluate the thickness of deposition layers in the past, the measurement area is still limited, and it is difficult to extend the measurement to a wide area in a vacuum vessel. In addition, when analyzing a shiny object such as metal, the reflected light from the specular reflection is strong. Therefore, the position of the sensor is affected by this effect. Thus, the color analysis of the object may not be accurate due to the strong influence of the specular reflection. The earlier study of color analysis indicates that the measurement area was not sufficient, and the accurate reflection rate was difficult to be evaluated in the metal object.

In this chapter, we introduce an example of the application of the color analysis method to the LHD using a color analyzer, which can be utilized on the metal

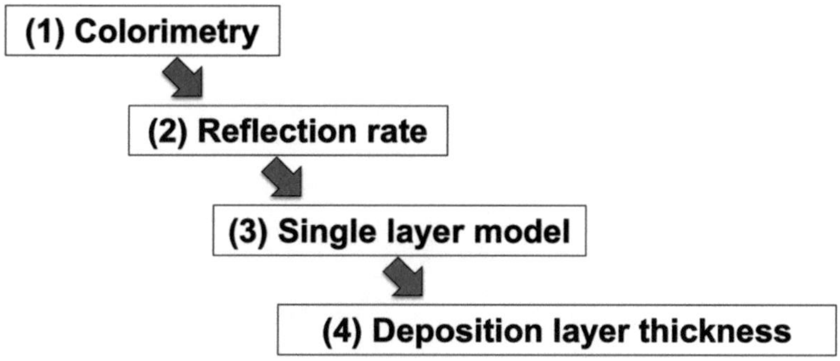

Figure 1.
The process from colour analysis to deposition layer thickness evaluation.

surface and evaluate the thickness of the deposition layer over a wide area. To evaluate the thickness of the deposition layer by color analysis, four processes are carried out, as shown in **Figure 1**. The evaluation of the thickness of the deposition layer from the color analysis is a four-step process, as shown in **Figure 1**. Process 1 is a color analysis using a color analyzer. Process 2 shows that the color analysis measurement is equivalent to the reflection measurement from a physical point of view, and Process 3 shows that the relationship between the reflection and the thickness of the deposition layer can be explained by the single-layer model. Finally, in Process 4, the thickness of the deposition layer is evaluated from the results obtained from color analysis. In this chapter, each process is explained in detail.

3. Compact color analyzer

A compact color analyzer (model: DM-1) manufactured by Hitachi Metals in Japan was used to evaluate the thickness of a wide range of deposition layers on the first wall of the LHD vacuum vessel [8]. A photograph of the color analyzer is shown in **Figure 2(a)**. The main feature of this analyzer is that it has an LED integrating sphere inside the analyzer. The internal structure of the integrating sphere is shown in **Figure 2(b)**. The light emitted from the LED is injected into the object as a homogeneous standard light by a diffuser. After that, the photodiode sensor captures the light reflected from the object, which includes both positive reflection and diffusive light. Then, numerical values output as the intensity of three specific types of light, so-called R, G, B (Red, Green, Blue), with central peaks at red (615 nm), green (540 nm) and blue (465 nm). This color analyzer can measure not only R, G and B but also hue, saturation and brightness. The specifications of the color analyzer are given in **Table 1**.

The color analyzer is lightweight, small in size, and user-friendly, so that it can be easily carried into the vacuum vessel opened to the atmosphere after the plasma experiment. In addition, the measurement time is only about 3 seconds, which makes it possible to measure many points in a short time. Furthermore, the internal memory enables continuous data storage, and the rechargeable battery eliminates the need for a continuous AC power supply. Here, to evaluate the accuracy of the color analyzer, we calibrated it using about 400 color sample books (DIC Color Guide, 19th Edition, PART1, 3) whose R, G and B values were known in advance [9]. A summary of the calibration results is shown below; the R, G and B values have an offset, and their values are all similar. The R, G and B values have similar characteristics of sensitivity with high sensitivity in large values and low sensitivity in small values. A high RGB value indicates that the image is close to white, and

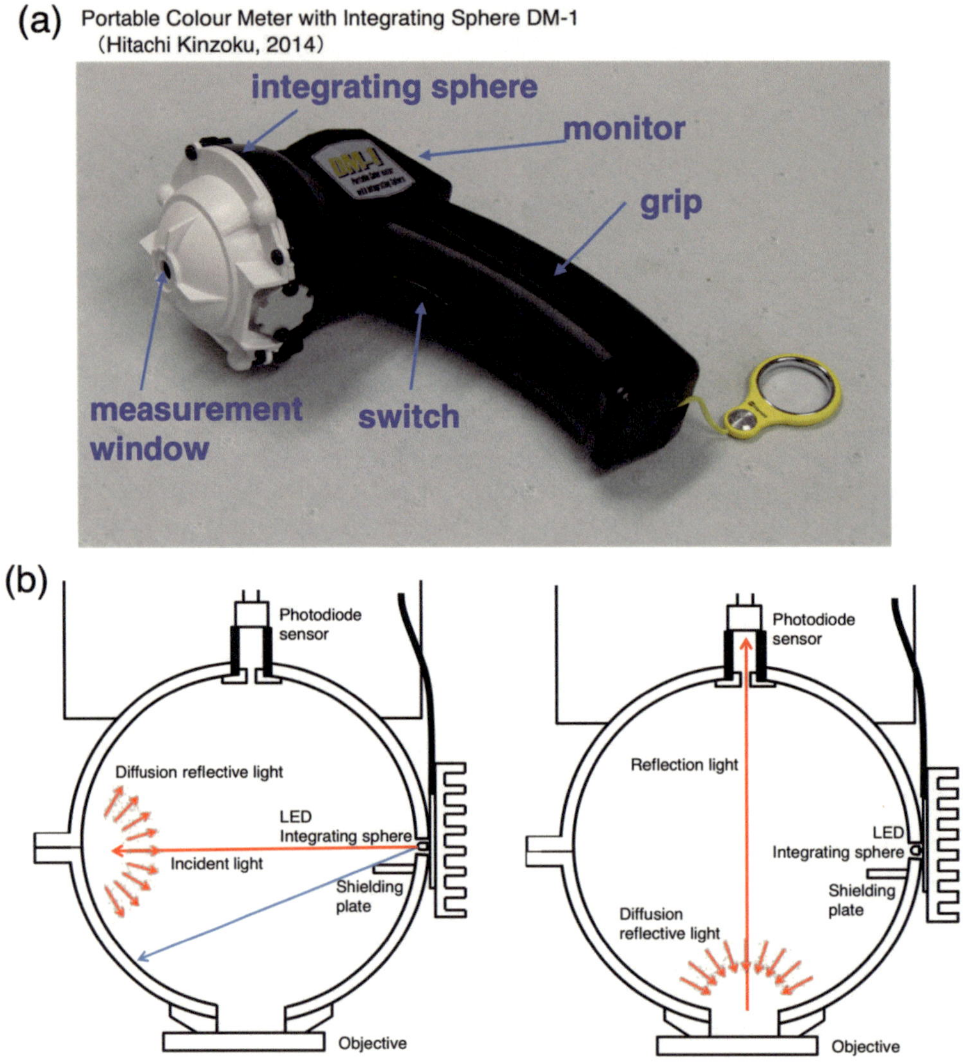

Figure 2.
(a) Photograph of the colour analyzer (model: DM-1) and (b) Operation principle of colorimetry method. Formation of incident light from the integrating sphere and capture of light reflected from the target [10].

Measurement	RGB (Red, Green, Blue) HSV (Hue, Saturation, Value)
Measurement window diameter	Φ8.1 mm
Internal diameter of integrating sphere;	Φ47 mm
Light source	White LED
RGB range	0~1023
Weight	~160 g
Measurement time	3 seconds
Record	USB memory installed in analyzer

Table 1.
Specifications of the colour analyzer.

a low RGB value indicates that the image is close to black, i.e., the image is more sensitive when the color is white and less sensitive when the color is black. This is similar to the human eye in general, which finds it more difficult to distinguish the difference in color between black than white. From the viewpoint of the evaluation of the deposition layer, it is desirable to have sensitivity up to the region of low RGB value, and further development of the color analyzer is expected in the future.

4. Experimental results

4.1 Relation between reflection rate and RGB values measured by the color analyzer

In this section, the physical meaning of the RGB values measured by the color analyzer is considered. As described in Section 2, the value measured by the color analyzer expresses the intensity of the reflected light from an object by spectroscopic RGB values of specific wavelengths in visible light. The correlation between the reflection rate and the RGB values was investigated using long-term samples irradiated by plasma, installed in the vacuum vessel of the LHD. The long-term irradiated samples are made of substrates of stainless steel and are placed on the first wall of the toroidal section at various locations from the inside to the outside of the torus. They are exposed to the plasma during one experimental cycle (see **Figure 6(b)** for the locations). The optical reflection spectra of these samples, which were removed from the vacuum vessel after the plasma experiment, were evaluated using spectroscopic ellipsometry. At the same time, the RGB values of the samples were measured with a color analyzer. **Figure 3** shows the relationship between the RGB values and the reflection rate. The reflection rates were obtained by averaging the values at the

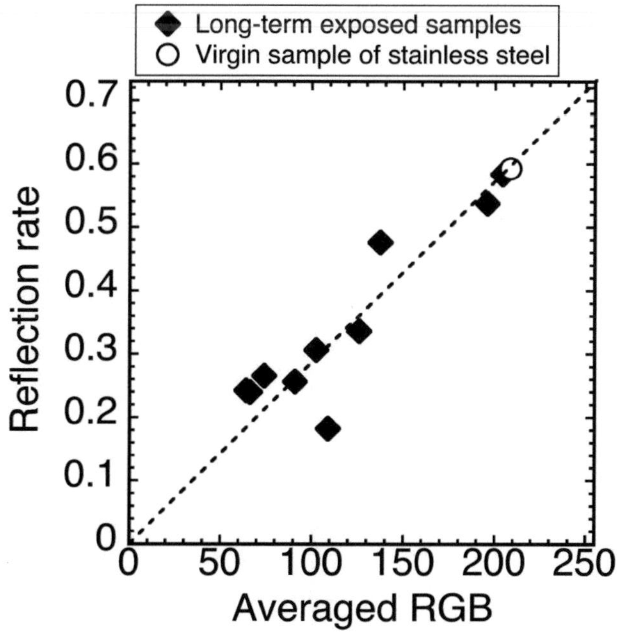

Figure 3.
The relationship between the RGB values measured by the colour analyzer and the reflectance measured by the spectroscopic ellipsometer [11].

wavelengths of red (615 nm), green (540 nm) and blue (465 nm) from the optical reflection spectra. For reference, the unexposed samples are plotted on the same graph. The reflection rate and the RGB values show a linear relationship, indicating that the RGB values measured by the color analyzer express the reflection rate. The surface roughness of the sample may also affect the RGB values. However, the cross-section of the sample observed by Transmission Electron Microscopy (TEM) shows that the surface roughness is less than 100 nm, which is lower than the wavelength of visible light, so the effect of surface roughness is considered to be small.

4.2 Evaluation of the thickness of the deposition layer by TEM cross-sectional observation

In the previous section, it was confirmed that the RGB values measured by the color analyzer are equivalent to the reflection rate. To estimate the thickness of the deposition layer from the reflection rate, it is necessary to clarify the relationship between them. The cross-sections of the long-term installation samples were cut by the Focus Ion Beam (FIB), and the thickness of the deposition layer of the long-term irradiated samples was evaluated by TEM observation. **Figure 4** shows the TEM images of each sample. In some samples, it is difficult to identify the interface of the deposition layer, and tungsten is deposited on the surface of the sample to prevent surface damage during FIB cutting. Although, the TEM images show interesting features such as directional structures in the deposition layer and blistering of the substrate, we will only focus on the thickness of the deposition layer in this chapter. **Figure 5** shows the relationship between the thickness of the deposition layer, evaluated from the TEM images, and the reflection rate measured by the color analyzer and spectroscopic ellipsometer.

4.3 Relationship between the experimental results and the single-layer model

To discuss the relationship between the thickness of the deposition layer and the reflection rate, we consider a single-layer model. In the model, we assume a simple

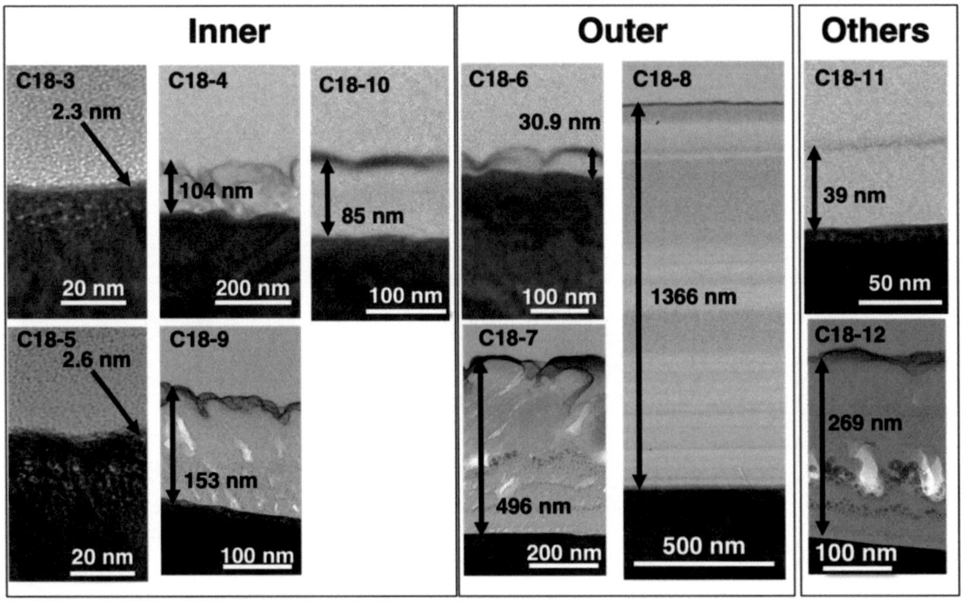

Figure 4.
TEM cross-sectional images of long-term irradiated samples [11].

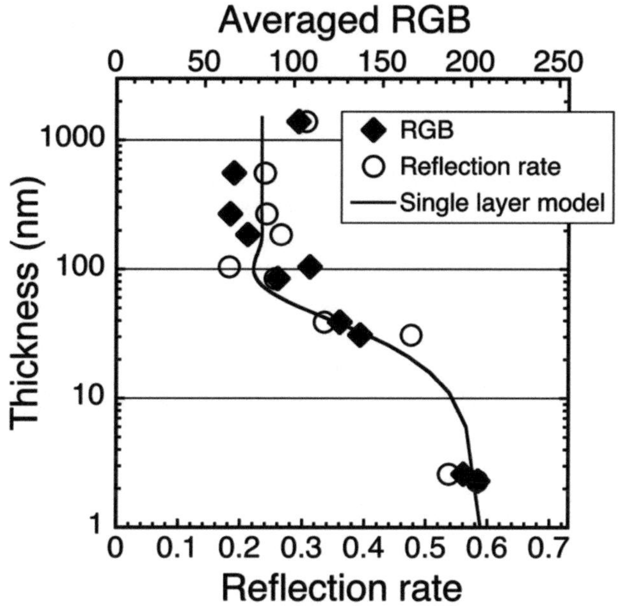

Figure 5.
Relationship between thickness of deposition layer and reflection rate using single layer model [11].

three-layers: the atmospheric layer, the deposition layer and the substrate layer. The reflection rate, R_{ref}, can be expressed as follows [12].

$$\varnothing = \frac{2\pi N_f \cos\theta}{\lambda}, \tag{2}$$

$$r = \frac{r_0 + r_1 \exp(i2\varnothing)}{1 + r_1 r_0 \exp(i2\varnothing)}, \tag{3}$$

$$R_{ref} = |r|^2 \tag{4}$$

where ϕ, λ and θ are the phase difference, the wavelength of the incident light and the angle of incidence of the light, respectively. N_f and d are the refractive index and thickness of the deposited layer, r is the electric field ratio of the reflected light to the incident light, and r_0 and r_1 are the Fresnel reflection coefficients at the air-deposition layer boundary and the deposition layer-substrate layer boundary, respectively. This simple model shows that the reflection rate is nonlinearly dependent on the thickness of the deposition layer. For the polarization of the light, the ratio of S-wave to P-wave is assumed to be 1:1. The refractive index of the deposition layer was set to $n = 1.24$ and $k = 0.98$ based on the results of ellipsometric measurements. The refractive index of the stainless steel substrate is assumed to be $n = 1.5$, $k = 2.9$. The relationship between the thickness of the deposition layer and the reflection rate of the single-layer model is shown by the solid line in **Figure 5**. A clear dependence of the reflection rate on the thickness can be observed in the range of 10 nm to 100 nm. We now look at the thickness dependence of the reflection rate of the single-layer model. While there is a dependence of the reflection rate between 10 nm and 100 nm, the dependence becomes weaker when the thickness of the deposition layer is below 10 nm or above 100 nm. This may be because the reflection rate of the substrate dominates for the thin layer and the reflection rate of the deposition layer dominates for the thicker layer. The dependence of the single-layer model is similar to the experimental results of the reflection rate using a color analyzer and spectroscopic ellipsometry. Therefore, the relationship between the reflection rate and the thickness of the deposition layer can be explained by the single-layer model.

4.4 Evaluation of reflection rate distribution and thickness distribution of deposition layer on the first wall of helical coils

The reflection rate of the first wall on the helical coil in the same toroidal section, where the long-term irradiated samples described in 3.2 were installed, was measured using a color analyzer after the opening to the atmosphere in the vacuum vessel. The measured number of stainless steel protection plates was 530. The reproducibility of the measurements was confirmed by performing the measurements twice. **Figure 6(a)** shows the results of the reflection rate

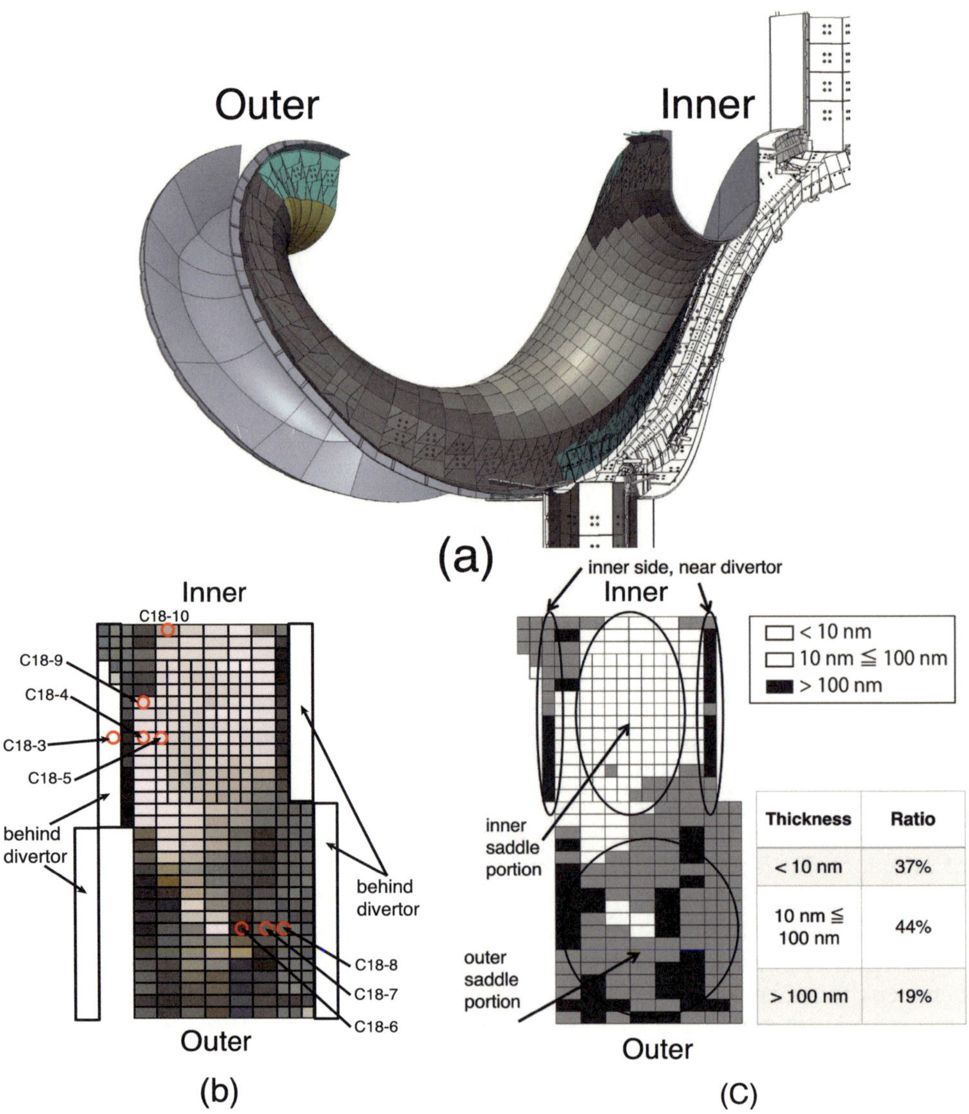

Figure 6.
(a) CAD showing the measured reflection rate, (b) developed view, and (c) deposition layer thickness distribution evaluated from reflection measurements [11].

measurements [11]. On the outside of the torus, the RGB values of most of the protection plates are low and the reflection rate is low. On the other hand, inside the torus, the RGB values are high, and the reflection rate is high except near the divertor plate. For clarity, **Figure 6(b)** shows the development figure of the measured stainless steel plates. These results suggest that deposition is dominant outside the torus, while erosion is dominant inside the torus. The reflection rate of the deposition layer is determined by the balance between erosion and deposition processes, which depends on the distance from the plasma and the view from the divertor plate. The RGB values are higher when the stainless steel plate is closer to the plasma and lower when it is farther from the plasma [9]. This relationship of the distance between the protection plate and the plasma with the reflection rate is interesting (for details, refer to [9]). Using the reflection rate measured by the color analyzer, and assuming that the thickness of the deposition layer follows the single-layer model, the thickness distribution of the deposition layer on the helical coil was evaluated. **Figure 6(c)** shows the thickness distribution of the deposition layer evaluated from the reflection rate measurement. As predicted from the reflection rate measurements, deposition is dominant on the outside of the torus, while erosion is dominant on the inside of the torus except near the divertor plate. In terms of the distribution of the deposition layers, 37% of the layers were 10 nm thick, 44% were 10–100 nm thick, and 19% were thicker than 100 nm. 60% of the first wall of the helical coil in the toroidal section was covered by the deposition layer. This result suggests that the region covered by the deposition layer shows an important role in the wall absorption of fuel particles. Quantitative comparison with global particle balance analysis is also performed using this experimental result, please refer to Ref [3].

Colorimetry has been applied to plasma experimental devices other than the LHD. Here we introduce an example of the application of colorimetry on the Wendelstein 7-X (W7-X) device at the Max-Planck-Institute for Plasma Physics in Germany [10]. Similar to the LHD, the W7-X is a large superconducting stellarator device [13]. **Figure 7** shows the results of colorimetry measurements in the W7-X on the surface of the panels covering the vacuum vessel and the evaluation of the deposition layers in different experimental campaigns. **Figure 7(a)** shows

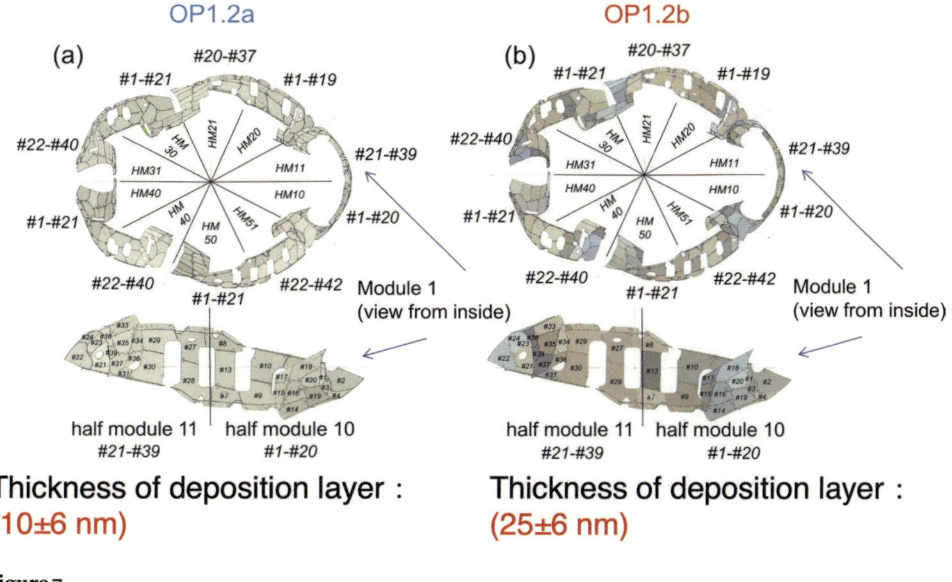

Figure 7.
Colour pattern of first wall panels after (a) OP1.2a and (b) OP1.2b in W7-X [10].

the estimated deposition layer after the experimental campaign of OP1.2a, where the average thickness of the deposition layer is predicted to be 10 ca. On the other hand, the deposition layer after the experimental campaign of OP1.2b is shown in **Figure 7(b)**. The thickness of the deposition layer is evaluated to be 25 ca. Thus, colorimetry provides useful information for discussing strategies for removing the deposition layer.

5. Summary

In this chapter, a color analysis method using a color analyzer is introduced, and it is shown that it is possible to evaluate the deposition layer formed on the plasma-facing wall by color analysis. As an example, the color analysis of the first wall on the helical coil of the LHD shows that the thickness distribution of the deposition layer on the inner and outer sides of the torus is different. Also, the different deposition layer thickness has been obtained in the different experimental campaigns in the W7-X. Although, color analysis is useful for surface analysis of plasma-facing walls, it may not be an in-situ diagnosis, because the current method requires a color analyzer to be brought into the vacuum vessel for measurement. However, using the principle of evaluating the thickness of the deposition layer from the color analysis, it could be applied as an in-situ diagnostic measurement. For example, if the color analyzer is held by a robot arm and operated remotely, in-situ diagnosis is possible. The color analysis method using the color analyzer has already been applied not only to the LHD and W7-X but also to the QUEST at Kyushu University and the GAMMA10 at Tsukuba University in Japan. In the future, color analysis will facilitate the understanding of the plasma-wall interaction in fusion plasma.

Acknowledgements

The author G.M. would like to express his gratitude to Professor Emeritus Naoaki Yoshida of Kyushu University, who allowed him to discuss color analysis methods and Dr. Kenji Matsumoto of Honda R&D Co., Ltd., who developed the color analyzer. G.M. also thanks Dr. Suguru Masuzaki of National Institute for Fusion Science, Dr. Chandra Prakash Dhard and Dr. habil. Dirk Naujoks of Max-Planck-Institute for Plasma Physics for their strong support of this study. This work was supported in part by the National Institute for Fusion Science under Grants (NIFSUMPP003-1 and NIFSULPP801) and by the NIFS Stellarator-Heliotron Association Committee (URSX209) and by JSPS KAKENHI grant (18H01203).

Abbreviations

He	helium
n	neutron
D	deuterium
T	tritium
LHD	Large Helical Device
NIFS	National Institute for Fusion Science
TEXTOR-94	Toroidal Experiment for Technology Oriented Research (TEXTOR)-94
ASDEX-U	Axially Symmetric Divertor Experiment Upgrade (ASDEX-U)
CCD	Charge Coupled Device

LED	Light Emitting Diode
RGB	Red, Green, Blue
AC	Alternating Current
TEM	Transmission Electron Microscopy
FIB	Focus Ion Beam
W7-X	Wendelstein 7-X
QUEST	"Q-shu" University Experiment with Steady-state spherical Tokamak
GAMMA10	a tandem mirror machine in Plasma Research Center at University of Tsukuba

Author details

Gen Motojima[1,2]

1 National Institute for Fusion Science, National Institutes of Natural Sciences, Gifu, Japan

2 Sokendai (The Graduate University for Advanced Studies), Gifu, Japan

*Address all correspondence to: motojima.gen@nifs.ac.jp

References

[1] Takeiri Y et al. Extension of the operational regime of the LHD towards a deuterium experiment. Nuclear Fusion. 2017;**57**:102023

[2] Motojima G et al. Global helium particle balance in LHD. Journal of Nuclear Materials. 2015;**463**:1080

[3] Motojima G et al. Global particle balance and its relationship with the plasma wall interaction emerging in long pulse discharges on the Large Helical Device. In: Proceeding for 26th IAEA Fusion Energy Conference, Kyoto, Japan: IAEA-FEC; 2016 EX/P8-3. Available from: https://conferences.iaea.org/event/98/contributions/11893/contribution.pdf

[4] Oya M et al. The role of the graphite divertor tiles in helium retention on the LHD wall. Nuclear Materials and Energy. 2017;**13**:58

[5] Tokitani M et al. Plasma wall interaction in long-pulse helium discharge in LHD–Microscopic modification of the wall surface and its impact on particle balance and impurity generation. Journal of Nuclear Materials. 2015;**463**:91

[6] Wienhold P et al. Time resolved observation of the erosion of boron containing protective coatings on wall elements of TEXTOR-94 by means of colorimetry. Journal of Nuclear Materials. 1997;**241-243**:804

[7] Pugno R et al. Investigation of local carbon transport in the ASDEX Upgrade divertor using $^{13}CH_4$ puffing. Journal of Nuclear Materials. 2009;**390 391**:68

[8] Available from: http://www.hitachikinzoku.info

[9] Born M, Wolf E. Principles of Optics 6th Edition Electromagnetic Theory of Propagation, Interference and Diffraction of Light. Oxford: Pergammon; 1980

[10] Motojima G et al. In-vessel colorimetry of Wendelstein 7-X first wall components: Variation of layer deposition distribution in OP1.2a and OP1.2b. Physica Scripta. 2020;**T171**: 014054

[11] Motojima G et al. Wide-range evaluation of the deposition layer thickness distribution on the first wall by reflection coefficient measurements. Nuclear Materials and Energy. 2017;**12**: 1219-1223

[12] Motojima G et al. Preliminary examination of reflection coefficient measurement of RGB lights on the first wall in LHD. Plasma and Fusion Research. 2015;**10**:1202074

[13] Klinger T et al. Overview of first Wendelstein 7-X high-performance operation. Nuclear Fusion. 2019;**59**: 112004

Chapter 9

Evaluation of Camouflage Coloration of Polyamide-6,6 Fabric by Comparing Simultaneous Spectrum in Visible and Near-Infrared Region for Defense Applications

Md. Anowar Hossain

Abstract

Polyamide-6,6 (PA-6,6) knitted fabric was coated with a complex combination of liquid-phase oxidized carbon black pigment (CBP) as light absorber and mono-sulfonated telon violet 3R (TVR) as acid dyes. Nitric acid (NA) moiety was used as liquid-phase oxidation of CBP and hydrophilic transformation of CBP-TVR. Thermoplastic polyurethane (TPU) and N, N-dimethylformamide (DMF) were formulated as cross-linker between composite mixture (CM) and PA-6,6 fabric. Six different CMs were coded for coating of PA-6, 6 fabric such as TPU-DMF, CBP-TPU-DMF, TVR-TPU-DMF, CBP-TVR-TPU-DMF, NA-TVR-TPU-DMF, NA-CBP-TVR-TPU-DMF. Structural, chromatic, and spectral reflection of CM coated PA-6,6 fabric was investigated by scanning electron microscopy, color measurement spectrophotometer, and Fourier transform infrared spectroscopy. CBP formulated PA-6,6 fabric was significantly remarked as maximum light absorber in both visible and near-infrared spectrum without allowing other parameters of treated PA-6,6 fabric. Therefore, minimum light reflection principle of CBP was indicated as camouflage material for camouflage textile coloration/finishing/patterning of simultaneous spectrum probe in visible and near-infrared spectrum. PA-6,6 fabric is very common fabrication for defense clothing, weapon, and vehicle netting against every combat background. This approach of simultaneous spectrum probe may be extended for concealment of target signature against high-performance defense surveillance.

Keywords: camouflage coloration, carbon black pigment, polyurethane, reflection, visible–near infrared

1. Introduction

Camouflage textiles are incorporated with clothing of special workers and multidimensional equipment in the vein of army, air force, navy, marines, coastal guards, paramilitary forces, uniforms for officers and soldiers in defense forces, tents, shelters and sheets, sandbags, flak jackets, helmets, camouflaged combat and flight uniforms, covering airplanes, guns and boats, creating deceptions, making armored vehicles, and so on [1]. Key parameters of camouflage coloration are spectral bandwidth of target signature, spectral characteristics of combat background, and reflectivity [2, 3]. Reflection of pigment formulation must match with combat background for design of high-performance camouflage textiles. Polyamide and polyester fabric coloration were experimented with carbon black pigment (CBP). The camouflage effect of CBP was recommended in near infrared (NIR), 1000–1200 nm [4]. Camouflage coloration was trialed with CBP for formulation of colorant on synthetic and regenerated textiles [5]. Visible-camouflage coloration was also investigated with pigment, reactive and vat dyes. Combination of CBP with other synthetic colorant depicted more prominent hue of black [6, 7]. CBP was used for camouflage coloration of polyethylene terephthalate textiles in comparison with natural brown hue against desert combat background. Hence, CBP property was also suggested for NIR camouflage [8]. In this experiment, reflection of CBP was compared with a synthetic dyes telon violet 3R (TVR).

1.1 Properties of CBP for suitability of technical coloration

CBP has versatile industrial applications as a pigment, which imparts a black hue [9, 10]. CBP is the most suitable for defense textiles due to having maximum properties of protection textiles. CBP has weatherability, chemical resistance [11, 12], abrasion resistance [13], electroconductivity [11], alkali resistance, light fastness [7], hydrophobicity [14], which are major considerations for coloration of defense textiles. CBP has ester bond, thioester bond, amide bond, amino bond, carbonyl bond, thiocarbonyl bond, sulfonyl bond [12], aldehyde bond, hydroxyl bond, hydrogen bond [9]. Crystalline CBP shows a high order of reactivity [15]. Quantitatively more application of CBP is found in the rubber industry to accelerate its mechanical properties, such as resistance to abrasion. CBP is generally termed as active carbon for improving the mechanical properties of rubber. CBP is widely used for black coloration in paints, varnishes, carbon paper, ribbons, printing inks [13, 16, 17]. Plastic articles, metal articles, wood, paper, inorganic materials [11], and polyamide fabric coloration with 70% nitric acid (NA) [18]. Oxidization process was used to improve the coloring property of printing inks, paint, and black coloration [13, 16, 17]. Due to hydrophobicity of CBP, it is difficult to apply textile substances. So CBP needs to be oxidized for improving jetness by chromic acid, ozone, hydrogen peroxide, sodium hypochloride, potassium permanganate, NA [14, 19] and sulfuric acid [7, 14]. CBP was grafted with hydrophilic polymeric monomer (alkali or ammonium carboxylate) for improving water compatibility and dispersibility [20]. CBP has dusting tendency [21], it is combined with the carrier to impart chromatic hue. CBP constituted generally as a pigment [22]. CBP with acid surface groupings is particularly suitable with binders [23]. Particle size, surface structure, surface size, pH, and dispersibility of CBP need to be considered for every polymeric binding system [24–26]. Diameter and nigrometer index of CBP critically modify the color of CBP. CBP shows high color for diameter 13 and 15 nm and nigrometer index 63 and 68, respectively. CBP depicts medium color for diameter 17, 20 nm and nigrometer index 71, 76. CBP shows regular color for diameter 25 nm and nigrometer index 80 [11]. Different categories of CBP are fisher lampblack (particle size 44 nm, p^H 7, carbon 96.7%, oxygen 0.9%, nitrogen 0.0%, sulfur 1.5%, hydrogen 0.6%), degussa special black (particle size 20 nm, pH 7, carbon 85.6%, oxygen 13.1%, nitrogen 0.3%, sulfur 0.4%, hydrogen 0.6%) and degussa color black FW 200 (particle size 13 nm, pH: 7, carbon 79.2%, oxygen 19.3%, nitrogen

0.4%, sulfur 0.4%, hydrogen 0.7%) [27]. Natural gas and hydrocarbon are the raw materials of CBP [28]. Coarser particle of CBP shows lower depth of color, and finer particle shows higher depth of color [23] due to its difference in scattering values with difference in size and surface area. Hence CBP can be applied for chromatic modification of textile substances such as PA-6,6, cotton, wool, and acrylic fabric.

1.2 Properties of TPU for suitability of textile coating

Thermoplastic polyurethane (TPU) elastomers are a cross-linking agent having low density and high resiliency [29], dispersibility, and polymeric film forming ability was considered the criterion for applying in acid medium coating. TPU is also applicable for printing paste formulation of synthetic fabric coloration [30]. TPU creates capsule shell of CBP [31]. TPU was formulated for the dispersion of pigment [18]. TPU has structural form of hydroxy-terminated polybutadiene and polyether-ester based prepolymers with diisocyanates such as hexamethylene diisocyanate [32]. Chemical reaction between polyols and polyisocyanates compounds existing hydroxyl groups reacts with isocyanates [18]. Therefore, structural features and properties of TPU are suitable for textile coating, textile finishing and printing.

2. Technical approach of camouflage textiles formulation with CBP

Reviewed patent and publications of CBP-based coloration exhibit that CBP has enumerated applications of coloring agent but limited applications on textile coloration. Simultaneous spectrum probe camouflage coloration is almost new concept in camouflage engineering. CBP has harmonized with the demanded textile properties for defense textiles such as low reflection, weatherability, chemical resistance, good rubbing fastness, and light-fastness properties, etc. Light scattering/absorbing principle of CBP materials have been predicted the possibility of deceiving reflection against combat background. CBP has multidimensional color forming functional groups such as carbonyl group, amide group, carboxyl group, which can deceive the surrounding color. The color tone of this functional group differs in spectral responses in Vis–NIR absorbance or reflectance-based spectrophotometric color evaluation and Vis–NIR spectral responses against wavelength. CBP has possibility to modify the hue of other color forming functional group. Light falling on CBP treated fabric diffuses or light is absorbed by the CBP influencing the chromatic hue/hiding the color hue. Hence the chromatic behavior of CBP on textile surface and its moiety with other colorant need to be identified for camouflage coloration. Color combination with CBP may conceal the chromatic hue of red, green, yellow, blue. CBP can be applied on textile substances for special worker's clothing like camouflage textiles as defense wear or protection tents for special requirements. There is very limited research on CBP formulated properties introduced for augmentation of camouflage textiles. Camouflage textiles can be identified by illumination properties of chromatic hue by spectral responses in both visible and NIR ranges as the chromatic hue in Vis and NIR range are not same. The reflectance can be identified by simultaneous spectrum in Vis and NIR region easily by using VIS–NIR/UV–Vis/UV–Vis–NIR reflectance spectroscopy.

3. Materials

130 GSM PA-6,6 knitted fabric collected from local supplier was used for experimentation. Water-insoluble and laboratory-grade CBP was used for PA-6,6 fabric

coloration. TPU-Texalon 598 A in tiny pellet, white color in appearance was selected as cross-linking agent. N, N-dimethylformamide (DMF) was used as solvent for making a viscous solution of TPU. Laboratory-grade NA (70%) moiety was used for liquid-phase oxidation of CBP surface, functioning as hydrophilic vehicle of CBP-TVR into PA-6,6 and overall enhancement of rheopectic property in composite mixture (CM). The components of CM were used without further purification as received from supplier. Mathis laboratory hot plate, Werser Mathis AG, Rutisbergstrasse 3, CH-8156, Oberhasli/Zurich, Switzerland, automatic temperature controllable dryer, and a hand-operated coating processor with 2 mm roller were used for camouflage coloration.

3.1 Structure of CBP-TVR-TPU chemical compounds formulated for experimentation

The chemical structure of CBP [9], TVR [33], TPU compounds [29], and PA-6,6 fabric [34] have been shown below in **Figures 1–4**.

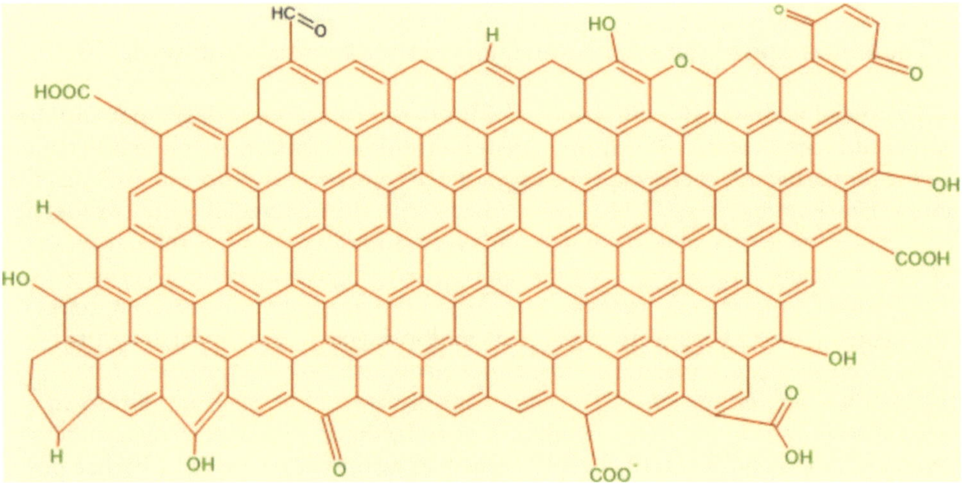

Figure 1.
Chemical structure of CBP used in this experimentation.

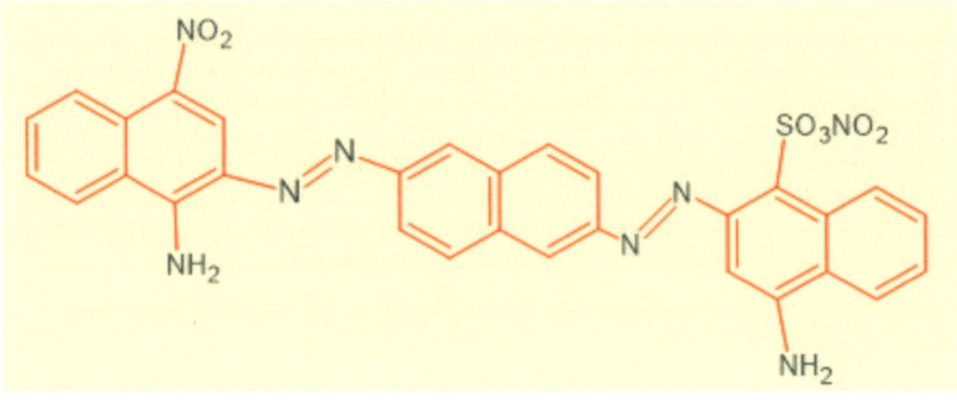

Figure 2.
Structure of TVR used in this experimentation.

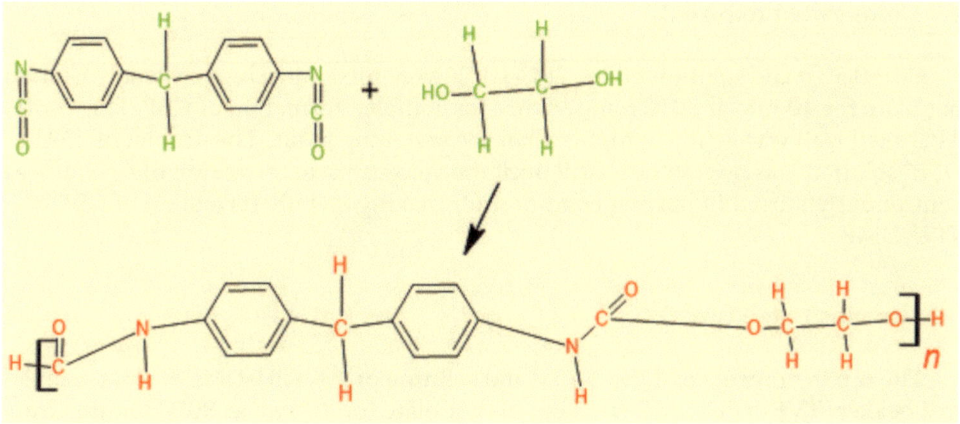

Figure 3.
Polymeric structure of TPU cross-linker used in this experimentation.

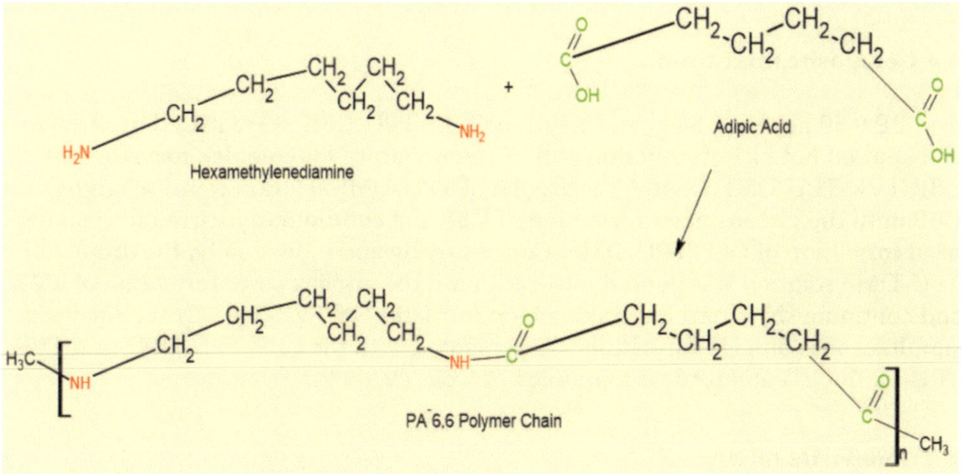

Figure 4.
Polymeric structure of PA-6,6 used in this experimentation.

4. Methods and preparation

Solution of TPU-DMF and its other variant of five more composite-mixed-complex formation of TPU-DMF were categorized and coded by CM-01 (TPU-DMF), CM-02 (CBP-TPU-DMF), CM-03 (TVR-TPU-DMF), CM-04 (CBP-TVR-TPU-DMF), CM-05 (NA-TVR-TPU-DMF), and CM-06 (CBP-NA-TVR-TPU-DMF).

4.1 Composite mixture-1

Tiny pellet of TPU (3%) was solubilized into DMF to form TPU-DMF solution, using a mini magnetic stirrer and an automatic shaker for 10 days in room temperature. The stirring and shaking periods were repeatedly checked and waited for maximum label of solubilization of TPU. The extended time was used for highest level of solubilization as no additional testing was implemented for this process of solubilization. TPU-DMF solution was added for each complex formation of CBP-TVR-NA.

4.2 Composite mixture-2

CBP 0.50 g and solution of 3% TPU-DMF were mixed in a beaker. It was kept on hot plate for 40 min at 80°C temperature for complex formation of CBP-TPU-DMF. TPU shell wall was kept just higher than the softening point. The droplet of TPU-DMF solution was poured into CBP until the viscous paste formation of CBP. It was continuously stirred in heating condition for making globule formation of CBP-TPU-DMF.

4.3 Composite mixture-3

The mixing process of TVR, 1.50 g and solution of 3% TPU-DMF were continued in a beaker. TVR-TPU-DMF was kept on hot plate for 40 min at 80°C temperature for complex formation. The droplet stirring process was followed for making bead formation of TVR-TPU-DMF. The droplet of TPU-DMF solution was poured into TVR until the viscous paste formation of TVR and continuously stirred for making bead formation of TVR-TPU-DMF.

4.4 Composite mixture-4

CBP 0.50 g, TVR 1.50 g, and solution of 3% TPU-DMF were mixed in a beaker and kept on hot plate for 40 min at 80°C temperature for complex formation of CBP-TVR-TPU-DMF. Firstly, the droplet of TPU-DMF solution was discharged into CBP until the viscous paste formation of CBP and continuously stirred for making bead formation of CBP-TPU-DMF. Consecutively and individually, the droplet of TPU-DMF solution was poured into TVR until the viscous paste formation of TVR and continuously stirred for making blob formation of TVR-TPU-DMF. Finally, a simultaneous complex formation was generated with the CBP-TPU-DMF and TVR-TPU-DMF (1:1) and formed a complex of CBP-TVR-TPU-DMF.

4.5 Composite mixture-5

TVR 1.50 g, NA (50%) and solution of 3% TPU-DMF were combined in beaker and heated for 40 min at 80°C temperature for complex formation of NA-TVR--TPU-DMF. The droplet of TPU-DMF and NA solution was poured into TVR until the viscous paste formation of TVR and continuously stirred for making bead formation of NA-TVR-TPU-DMF.

4.6 Composite mixture-6

CBP 0.50 g, TVR 1.50 g, NA (50%), and solution of 3% TPU-DMF were formed as paste in a beaker and kept on hot plate for 40 min at 80°C temperature for complex formation of CBP-NA-TVR-TPU-DMF. Firstly, the droplet of NA solution and the droplet of TPU-DMF (1:1) were poured into CBP for liquid phase oxidation of CBP and bead formation of CBP-TPU-DMF. The process was continued until the viscous paste formation, continuously stirred for making bead formation of CBP-TPU-DMF. Sequentially and separately, the droplet of TPU-DMF-NA was poured into TVR until the viscous paste formation of TVR and continuously stirred for making bead formation of NA-TVR-TPU-DMF. Finally, the paste of CBP-TPU-DMF and NA-TVR-TPU-DMF were combined for complex formation of CBP-NA-TVR-TPU-DMF.

4.7 Coating on PA-6,6 fabric

PA-6,6 fabric was cut into required sizes (width-3inch × length-6inch) was uncontaminated with dipping in deionized water, then dried 50 min at 80°C in a heating chamber, and then the fabric was relaxed and cooled for 30 min at room temperature. Back part of PA-6,6 fabric was wrapped with aluminum foil paper, having thickness less than 0.2 mm for creating the artificial plain surface on the backside of fabric. This method can be repeated and applied for different types of plain roller surface of industrial coating machine. The fabric ends were tightly attached and laid on coating plate by thin adhesive paper and then the coating roller was used for hand coating system. This method was followed repeatedly for three-stroke coating process for even dispersion of TPU-DMF and its other variant of five more composite-mixed-complex mixture on fabric surface. Decontaminated PA-6,6 was coated with TPU-DMF, a three-stroke coating process following first stroke coating-by second stroke coating and then third stroke coating for even dispersion of TPU-DMF. Similar coating process was thus carried out with CM-02 (CBP-TPU-DMF), CM-03 (TVR-TPU-DMF), CM-04 (CBP-TVR-TPU-DMF), CM-05 (NA-TVR-TPU-DMF) and CM-06 (CBP-NA-TVR-TPU-DMF). Also, for further research experimentation purpose, sequential overlapped coating of one formulation over other was also carried out by coating sequentially CM-2 (CBP-TPU-DMF) in first stroke coating process, and then CM-03 (TVR-TPU-DMF) coating was made on it by second stroke coating and finally then CM-04 (CBP-TVR-TPU-DMF) was coated on it by third stroke coating process. Similarly, CM-02 (CBP-TPU-DMF) was coated by first stroke coating process and then CM-03 (TVR-TPU-DMF was coated by second stroke coating and finally CM-06 (CBP-NA-TVR-TPU-DMF) was sequentially coated by third stroke coating process. All the coated PA 6,6 fabrics were dried at 60°C for 60 min to proceed for testing.

5. Testing methods

5.1 Color measurement spectrophotometer

CIE, color parameters (L*, a*, b*) were measured by Hunter lab reflectance spectrophotometer, Color Flex EZ; model, 45/0 LAV; under testing conditions with geometry, 45°/0°; viewing area, large; D65 illumimnant/10°standard observer; room temperature, 18°C. This hunter lab illuminant spectrophotometer uses a xenon flash lamp to illuminate the test specimen, PA-6,6. The tonal variation of PA 6,6 fabric was subjected to colorimetric evaluation for determining CIE color coordinates L*, a*, b* values. The Hunter Lab reflectance spectrophotometer was calibrated in terms of highest darker and highest lighter by using standard black and white standard, for checking and matching standard values kept in machine software. To ensure sample opaqueness by minimizing the transparency of incident light; test specimen, fabric size (width-3inch × length 6-inch) was single folded in lengthwise (width 3-inch × length 3-inch) for placing on reflectance port.

5.2 Fourier transform infrared spectrometry

NIR scanning of treated and untreated PA-6,6 fabric was performed by FTIRS. A NIR background was standardized under diffuse reflection standard. Every sample is covered by sample cup to create reflection environment under a specified black standard, which is termed as "spectralon reflection." The sample port has sapphire/crystal window to capture reflection of sample. Machine-specified glass vial was

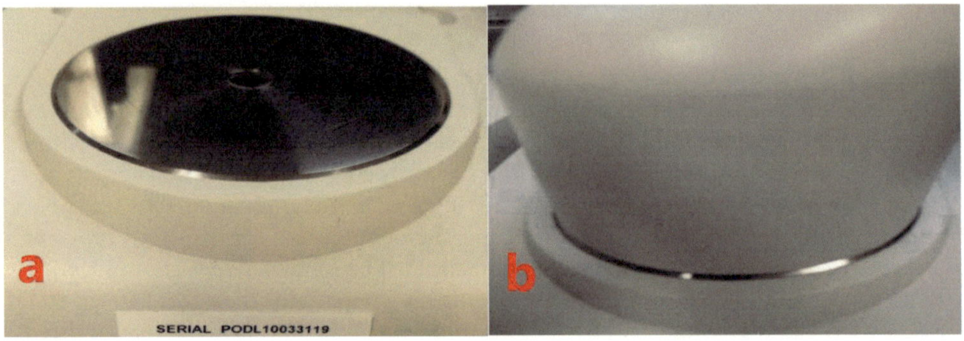

Figure 5.
Front view of FTIRS-NIR (a) uncovered diffuse sample port, (b) covered sample port with standardized black reference.

used for powder sample measurement of CBP-TVR. **Figure 5a** shows the sapphire window-sample port of FTIRS-NIR. **Figure 5b** shows the sample scanning condition under machine specified black standard.

5.3 Scanning electron microscopy (SEM)

Scanning electron microscopy (SEM) image was captured by TM 4000Plus model, Tabletop SEM, HITACHI, Japan. 15 KV electron acceleration was selected for every scanning of SEM. 100 magnification was performed for all image except NA treated fabric. NA-TPU-DMF and NA-TVR-TPU-DMF were captured at 25 magnifications to connote TPU effect in NA-CM-PA-6,6. Carbon conductive black tape was used for each sample mounting.

6. Results and discussion

6.1 Characterization and structural analysis by SEM

Figure 6, uncoated PA 6,6 fabric (a); TPU-DMF coated PA 6,6 fabric (b); CBP-TPU-DMF coated PA-6,6 fabric (c); TVR-TPU-DMF coated PA-6,6 fabric (d); CBP-TVR-TPU-DMF coated PA-6,6 fabric (e); NA-TVR-TPU-DMF coated PA-6,6 fabric (f); NA-CBP-TVR-TPU-DMF coated PA-6,6 fabric (g); blended CBP in water (h); blended CBP after 24 hour precipitation (i); blended CBP after 48 hour precipitation (j); 3% TPU in DMF (k), 3% TPU in DMF after addition of five drop NA; 5% of 70% NA (l); solidification slice of TPU-DMF in the presence of NA (l); SEM magnification of CBP particle (m) and its color thresholding structure of CBP particle (n); particle size of CBP (o); SEM magnification of TVR particle (p) and its color thresholding image of particle size (q); SEM magnification of PA 6,6 uncoated fabric (r); TPU-DMF coated PA 6,6 fabric (s) and its color thresholding image to refine TPU-DMF on fabric surface (t); CBP-TPU-DMF coated PA-6,6 fabric (u) and its color thresholding image to signify TPU-DMF on fabric surface (v); TVR-TPU-DMF coated PA-6,6 fabric (w) and its color thresholding image to identify TPU-DMF on fabric surface (x); CBP-TVR-TPU-DMF coated PA-6,6 fabric (y) and its color thresholding image to signify TPU-DMF on fabric surface (z); NA-TPU-DMF coated PA-6,6 fabric (z_1); NA-TVR-TPU-DMF coated PA-6,6 fabric (z_2); NA-CBP-TVR-TPU-DMF coated PA-6,6 fabric (z_3).

Figure 6 (i, j), water medium diluted CBP has been shown to signify the dispersibility of raw CBP. 0.5gm CBP was blended (30 minutes) with 150 ml

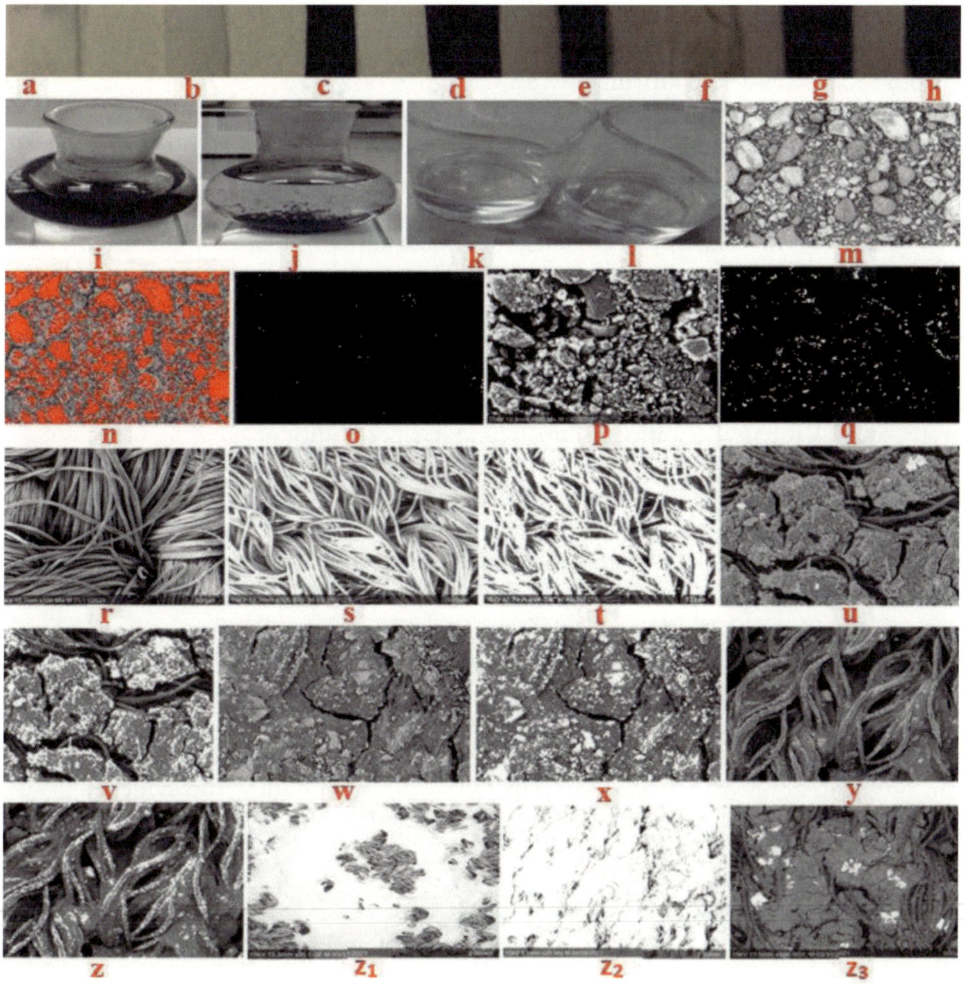

Figure 6.
Photograph of treated and untreated fabric (a-h), CBP and TPU-DMF derivative solution (i-l), SEM and color thresholding image of treated and untreated PA 6,6 fabric (m-z_3).

deionized water. The solution showed a precipitation/aggregation after keeping 24 and 48 hours in non-shaking condition. **Figure 6** (m) shows the structure of CBP. CBP particles are aggregated in different direction like a net structure which is demonstrated by red threshold, imageJ, **Figure 6** (n). The magnified cross section of CBP signifies the particle size, water insolubility, and randomly oriented micro-structure of surface for diffuse reflection due to surface roughness, **Figure 6** (m-q). Particle size of CBP denotes the higher degree of cracking on CBP coated PA-6,6 fabric rather than TVR coated surface, **Figure 6** (u-x). The existing of CBP-TPU-DMF has been clearly identified on PA 6,6 coated fabric, **Figure 6** (u, v, z_3). The cementing of CBP-TPU-DMF-PA-6,6 is higher in comparison with NA-CBP-TPU-PA-6,6. CBP-TPU formed a darker appearance due to higher prominence of black region on CBP-coated surface. CBP coating with TPU-DMF binder showed a cross-linking in CBP-TPU-DMF-PA-6,6. The functional group of CBP (-COOH) and the functional group of TPU (-NHCO) was formed a CBP-TPU matrix on the surface of PA-6,6 fabric. **Figure 6** (t, v, x, z, z_1, z_2, z_3), TPU was visualized in SEM image as whitish color surface. The prominence of TPU was recorded as minor when it was formed matrix with CBP. TPU binder was almost uniformly distributed on PA-6,6 fabric surface, **Figure 6** (t-z, z_3). TPU-CBP was implanted and well dispersed on

PA-6,6 surfaces. **Figure 6** (k, l), NA-TPU-DMF was also characterized in the presence of NA and TPU was showed freezing tendency. The application of TPU on PA-6,6 fabric was also confirmed by SEM, **Figure 6** (z_1, z_2).

6.1.1 Coupling of CM and SEM evidence of material suitability for visible-NIR camouflage

Figure 6 (o- z_3) shows the evidence of almost suitability of CBP material. The simultaneous applications of NA-TPU-CBP-TVR on PA-6,6 fabric surface have been confirmed by SEM image. The structural representation and coupling with similar functional groups have been revealed with individual coupling reactions with PA-6,6 **Figures 7–12**. Oxidized and aqueous CBP have amide group. PA-6,6 has also similar group, thus it may create a coupling of CBP-PA-6,6. CBP has amine group and TVR has also amine group. There is a possibility of bonding between CBP-TVR. Similarly, CBP has carbonyl group and TPU has also carbonyl group, thus CBP-TPU moiety may form in CBP modified PA-6,6 fabric. Carbonyl carbon of CBP may be bonded with two hydrogen atoms of TPU. Hence there is possibility of cross-linking. TPU may be adsorbed by CBP and dispersed the carbon black due to its structural formation of steric hindrance.

6.1.1.1 Coupling of TPU-DMF treated PA-6,6

Figure 6 (s, t) shows that polymeric coupling of PA-6,6-TPU, which are bonded with common functional group carbonyl (**Figure 7**). Structural resemblance of carbonyl and amine functional group in TPU-PA-6,6 may show the evidence of yellowish and greenish illumination on TPU treated PA-6,6 and without treated PA-6,6.

6.1.1.2 Coupling of CBP-TPU-DMF treated PA-6,6

Figure 6 (u, v) shows the coupling among PA-6,6-CBP-TPU. CBP shows multifunctional group including carboxyl, hydroxyl, aldehyde, carbonyl, ether, and it may have other functional group including color forming groups (**Figure 8**). CBP-TPU has amino group and carbonyl group. PA-6,6 has amino group and carbonyl

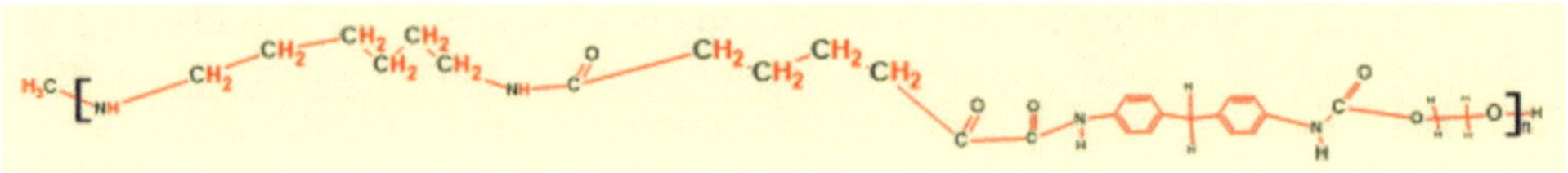

Figure 7.
Coupling of TPU-DMF treated PA-6,6.

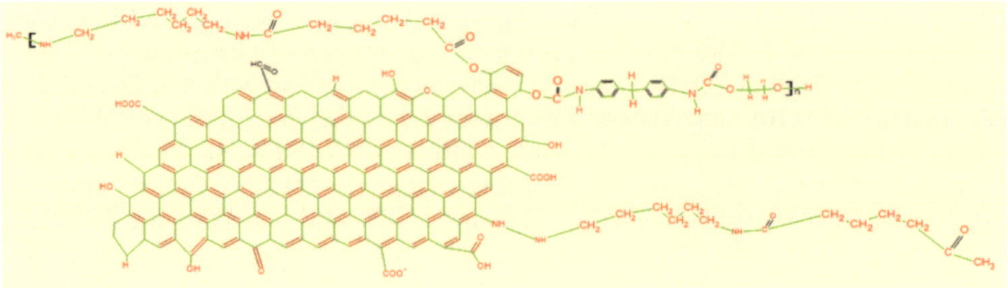

Figure 8.
Coupling of CBP-TPU-DMF treated PA-6,6.

group. Amino group and carbonyl group remaining PA-6,6 surface may generate reddish and yellowish hue but the reflectance of yellowish may be weakened and, it may be turned to bluish coordinates. So, the mechanism of tonal variation on textile surface/hiding of color hue may have occurred due to multidimensional color reflection of different hue as CBP has various numbers of functional groups for different color formation on same surface. Structure of CBP surface may also vibrate the camouflage concept of illumination due to spherical surface of CBP. COOH, carbonyl group, and OH group presence in CBP may reflect reddish hue. Amine group may be an indication of yellowish hue. Therefore, color combination of CBP may show darker hue by reflecting almost black hue. When any group with colorant of yellow, blue, and red are mixed with CBP, it may hide the chromatic intensity of added colorant and makes the dull shade to the observer due to remaining multifunctional and multicolored forming group. The multifunctional mechanism of hiding tendency of color hue of CBP may be functioning as key principle of camouflage coloration on textile substances.

6.1.1.3 Coupling of TVR-TPU-DMF treated PA-6,6

Figure 6 (w, x) shows the coupling of amide group among PA-6,6-TPU-TVR (**Figure 9**). TVR dyes has nitro group, amino group, and azo group. Nitro group may create blueish/yellowish hue, azo group may contain red/yellowish hue and amine group may prominent the yellowish reflection on the surface.

6.1.1.4 Coupling of CBP-TVR- TPU-DMF treated PA-6,6

Figure 6 (y, z) represents the NH coupling between CBP and PA-6,6 (**Figure 10**). Benzene coupling may occur between CBP and TVR. Correspondingly

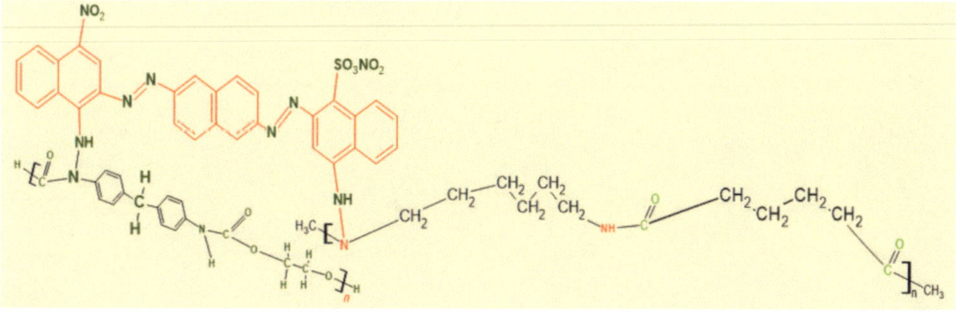

Figure 9.
Coupling of TVR-TPU-DMF treated PA-6,6.

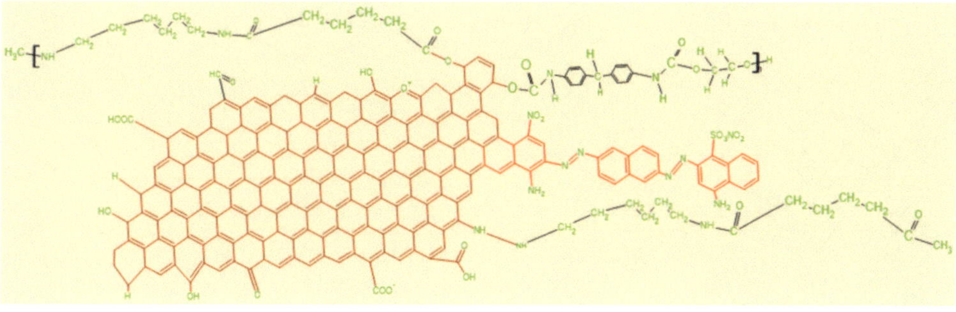

Figure 10.
Coupling of CBP-TVR- TPU-DMF treated PA-6,6.

carbonyl bonding is also evidenced between TPU and PA-6,6 due to structural view in SEM magnification. TVR dyes has nitro group, azo group and amino group. PA-6,6 and TPU has also amino group. Nitro group in TVR may create bluish and yellowish hue, azo group depicts red/yellowish hue and amine group may enhance the reflection of yellow hue. CBP and TVR has similarity in benzene group and amino group. TVR encompasses nitro group and azo group, which may reflect bluish and yellowish tone, azo group may illuminate red/yellowish hue and amine group may accelerate the yellow color reflection. Hydrophilic coupling between amine group of TVR and PA-6,6 may not be happened due to absence of hydrophilic vehicle NA. CBP has COOH, OH group, amino group, and carbonyl group, which may create chromatic change of color hue with another functional group.

6.1.1.5 Coupling of NA-TVR-TPU-DMF treated PA-6,6

Figure 6 (z_2) shows that amine group of TVR has related to both PA-6,6 and TPU, it seems the structural similarity (**Figure 11**). TVR has nitro group. Amino group of PA-6,6 and TPU may create a good color combination in NA-TVR-TPU-DMF when oxidized CBP has not been used in this process of coloration. For comparison, the reaction of oxidized carbon has been remarked with arrow sign connected with nitro group. In this coupling, there is no function of oxidized carbon due to missing of CBP in CM.

6.1.1.6 Coupling of NA-CBP-TVR-TPU-DMF treated PA-6,6

Figure 6 (z_3) shows amide linkage between CBP and PA-6,6 (**Figure 12**). Amine bonding may create between TVR and PA-6,6. Carbonyl bonding may form among CBP-TPU-PA-6,6. Oxidized carbon produces NO_2 and it has been showed with

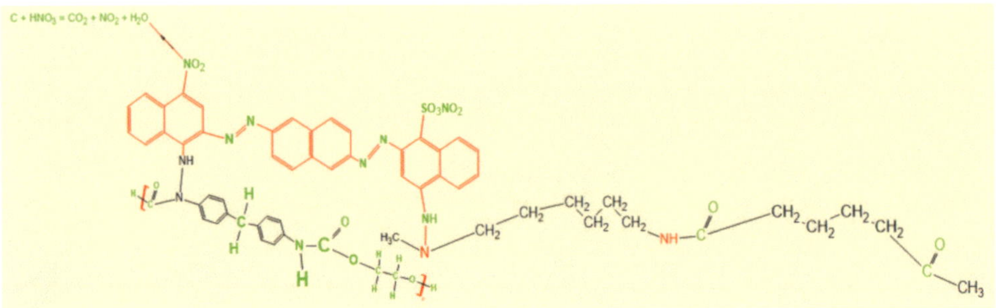

Figure 11.
Coupling of NA-TVR-TPU-DMF treated PA-6,6.

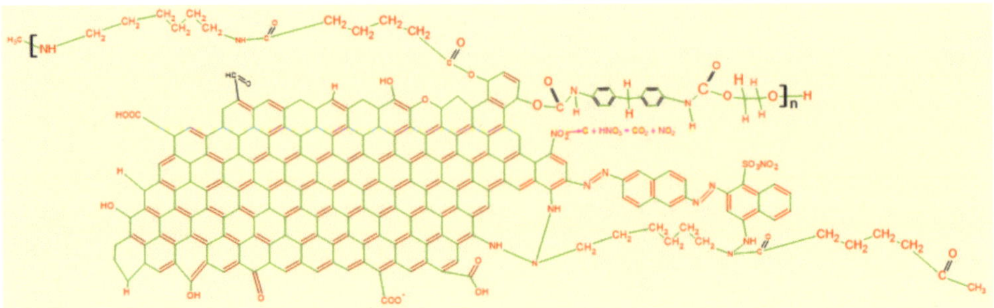

Figure 12.
Coupling of NA-CBP-TVR-TPU-DMF treated PA-6,6.

arrow marked to mention the similar group of TVR dyes (NO_2). The structure of oxidized carbon may be changed due to NA oxidation of CBP. The mechanism may influence the altering of reflection/wavelength in visible-NIR spectrum. The simultaneous combination of oxidized carbon and CBP modified PA-6,6 may replace the surface reflection significantly. The color forming functional group has been predicted the changing factors of chromatic hue. PA-6,6 has carbonyl group and amine group. TPU has also carbonyl group. Remaining unusual functional group of CBP like COOH, OH may reflect reddish hue, carbonyl group may reflect greenish/ yellowish hue and amine group may illuminate greenish/reddish hue. Spherical shape of CBP may hide the real illumination of nitro and azo group in TVR when nitro group may reflect bluish and yellowish tone and azo group may illuminate red/yellowish hue. Hence the actual prominence of hue is being decisively changed/ hided by the multifunctional tonal group of CBP.

6.2 Camouflage and chromatic analysis in visible spectrum

Established Beer–Lambert law [35] of optical absorption of material, A = εlc, A = optical absorption of materials, ε = absorptivity of the materials, l = optical path length, c = concentration of the materials. The Beer–Lambert law can be implemented for the optical absorption on PA-6,6 fabric surface in different wavelength. The variation of optical absorption of CBP can be predicted for the feasibility of camouflage coloration. Optical absorption on PA-6,6 fabric surface versus replacement of CIE color hue is proportionally related for camouflage coloration. Higher degree of optical absorption depicts the reduced intensity of chromatic hue. In this experimentation, A = optical absorption of PA-6,6 fabric surface, ε = absorptivity of CBP-TVR-TPU, I = optical path length in 400–700 nm and c = concentration of CM which are TPU-DMF, CBP-TPU-DMF, TVR-TPU-DMF, CBP-TVR-TPU-DMF, NA-TVR-TPU-DMF, NA-CBP-TVR-TPU-DMF. CIE optical absorption and reflection of chromatic hue have been studied in optical path length, 400–700 nm. The concept demonstrates that “more optical absorption on textile surface = less reflection on textile surface = altering the chromatic hue = deceiving the target detection”. Therefore, CBP can create chromatic variation on textile surface and confuse the target detection to the observer. But optical path length 400–700 nm and concentration of surface CBP materials can alter the chromatic behaviors of PA-6,6 fabric. Camouflage coloration was observed in complex formation of CBP for surface hue modification of PA 6,6 fabric. The color outcome was identified based on light absorption and reflection of CIE color hue, L^*, a^*, b^*. The reflection changing of CBP-NA-TVR-TPU-DMF treated PA-6,6 fabric was remarked in terms of CIE color coordinates. CBP-NA-TVR-TPU-DMF treated fabric showed comparatively less reflection of light and color hue hiding tendency.

6.2.1 Spectrophotometric color comparison of treated PA-6,6 and untreated PA-6,6

Figures 13–16 illustrate that CBP-TPU-DMF, CBP-TVR-TPU-DMF and NA-CBP-TVR-TPU-DMF treated PA-6,6 fabric modified the light absorption and intensity of chromatic hue in CIE L^*, a^*, b^* due to existence of CBP. The value of L^*, a^*, b^* represents higher light absorption and minor reflection of chromatic hue, which may be elucidated the reflection changing of CBP modified PA-6,6 surface. The moderation of chromatic hue intensity identifies camouflage categories of reflection in visible range due to having a minimum value of lightness (L^*), minimizing the value of red-green coordinates (a^*) and blue-yellow coordinates (b^*). **Figures 13–15** demonstrate that NA oxidized CBP shows maximum absorption of light than other CBP treated PA-6,6. NA oxidized NA-CBP-TVR-TPU-DMF reveals

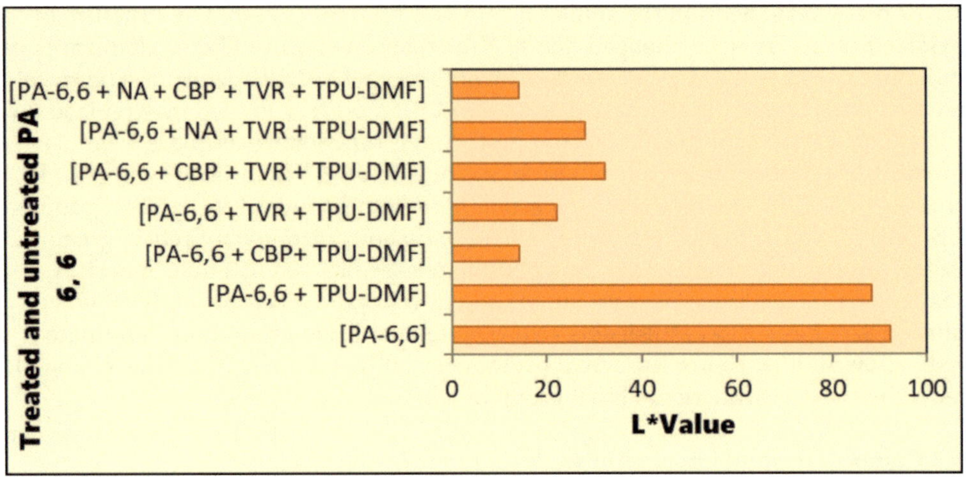

Figure 13.
CIE comparison of lightness hue (L) of treated and untreated PA-6,6 fabric.*

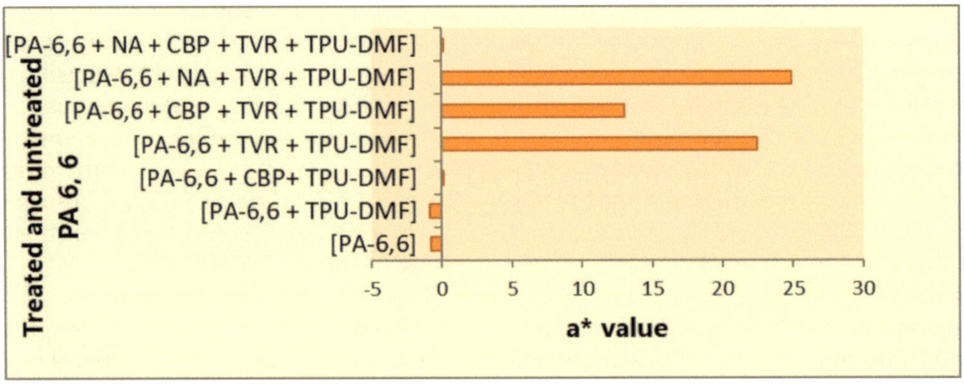

Figure 14.
CIE comparison of greenish/reddish hue (a) of treated and untreated PA-6,6 fabric.*

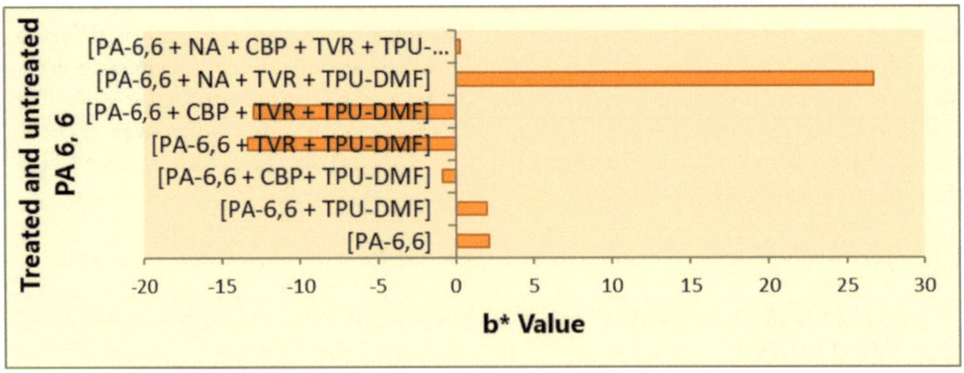

Figure 15.
CIE comparison of bluish/yellowish hue (b) of treated and untreated PA-6,6 fabric.*

CIE, L^* = 14.33, a^* = 0.15, b^* = 0.30 in wavelength, 400–700 nm. CBP is practically insoluble in water. In NA-CBP-TVR-DMF, NA is functioning to form hydrophilic transformation of CBP-TVR, as well as same time TPU-DMF is performing as artificial cross-linker between CM and PA-6,6 without influencing color reflection of any other component in CM. Thus, oxidized CBP molecules may be hided the sulfonated group of TVR dyes in NA-CBP-TVR-TPU-DMF mixture whereas CBP-TVR-TPU-DMF does not exhibit hiding of red and blue color hue due to absence of NA and lacking hydrophilic reaction. CM of CBP-TPU-DMF and NA-CBP-TVR-TPU-DMF exhibit nearest of neutral value in three-dimensional CIE color coordinates. CBP-TVR-DMF mixture reveals the positive value of a^* remaining reflection of reddish hue although theoretically the CM cannot make a complex due to absence of NA. Here, NA is not acting as hydrophilic vehicle of CBP in this CM. A proper reaction of TVR-CBP colorant may be held in between polyamide-6,6 and NA-CBP-TVR-TPU-DMF. In general theory of chromaticity, the color channel, $a^* = 0$ and $b^* = 0$ displays the true natural/neutral black color. The color wheel of NA-CBP-TVR-TPU-DMF shows $a^* = 0.15$, $b^* = 0.30$, therefore CBP has a propensity to neutral black. Similarly, NA free mixture of CBP-TPU-DMF and NA-CBP-TVR-TPU-DMF may prove minimum L^* value due to having higher absorption of light and very minimum reflection of light. Therefore, CBP modified PA-6,6 may create a deceiving environment of color hue to the observers. Thus, a neutral or minimizing tendency of a^* and b^* value can create the effect of hiding reddish hue which have an outcome of declining reflection of color molecules and increasing the absorbency of light. CBP modified PA-6,6 fabric surface may conceal the reflection of remaining TVR coloring molecule of red, green, blue, yellow. CBP treated surface may be decisively reduced the formation of cone shaped receptor for visible range of color vision. Rough surface of CBP treated PA-6,6 may generate the properties of diffuse reflection. D65 light source measurement of the reflectance characteristics of CBP is the key phenomenon of camouflage have been demonstrated in visible range 400–700 nm.

6.2.1.1 *Spectrophotometric color combination of untreated PA-6,6 and TPU-DMF-PA-6,6*

Figures 13–16; TPU-DMF treated PA-6,6 fabric and untreated PA-6,6 fabric was compared for identification of three-dimensional chromatic hue (L*, a*, b*) and its effects on PA-6,6 fabric surface. Spectrophotometric color combination of PA-6,6 shows greenish ($a^* = -0.79$) and yellowish ($b^* = 2.1$) hue, which are the nearest of

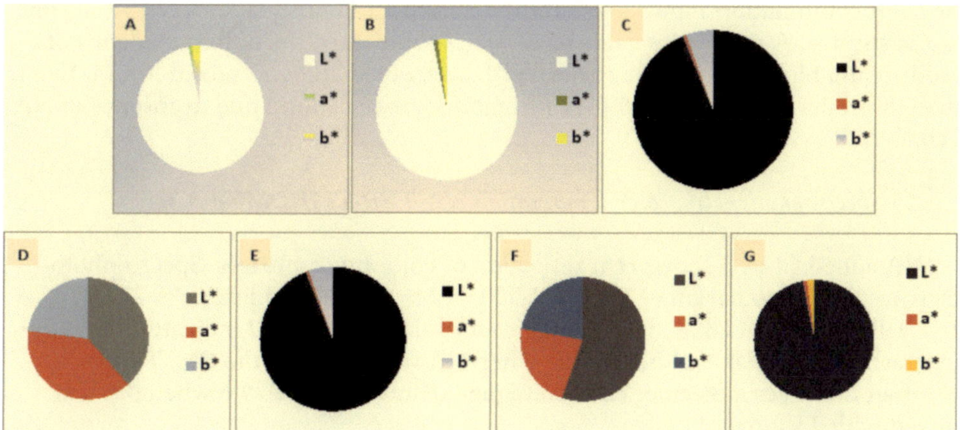

Figure 16.
Tristimulus intensity for variations and comparison of CIE L, a*, b* and representation of spectrophotometric color combination on a (PA-6,6), B (PA-6,6 + TPU-DMF), C (PA-6,6 + CBP+ TPU-DMF), D (PA-6,6 + TVR + TPU-DMF), E (PA-6,6 + CBP + TVR + TPU-DMF), F (PA-6,6 + NA + TVR + TPU-DMF), G (PA-6,6 + NA + CBP + TVR + TPU-DMF).*

neutral hue. A maximum value of L^* = 92.29 indicates almost white color in gray hue on fabric surface. Hence color combination of PA-6,6 indicates the maximum reflection of hue. Oppositely TPU-DMF treated PA-6,6 depicts greenish (a^* = −0.90) and yellowish (b^* = 1.96) hue, which also identify the nearest of neutral hue. A maximum value of L^* = 92.29 indicates almost white color/gray hue on the surface of fabric. Therefore, the color combination of TPU-DMF treated PA-6,6 also remarks the maximum reflection in CIE color coordinates. So, it can be compared that the reflection properties of CIE color hue are almost nearest/similarity in untreated PA-6,6 and TPU-DMF treated PA-6,6 fabric. It can be strongly said that TPU-DMF has minor effect on color changing. This study can be simply recommended for application of TPU-DMF for cross-linking of camouflage materials on textile surface.

6.2.1.2 Spectrophotometric color combination of CBP-TPU-DMF-PA-6,6

Complex application of CM-2 shows red (a^* = 0.08) and bluish (b^* = −0.84) hue which are the nearest of neutral hue. A minimum value of L^* = 14.21 have been pointed to almost black color tone on the surface of PA-6,6 fabric. CBP-TPU-DMF treated fabric shows less reflection of color hue. It may clarify gray hue, which is tending to black and hiding the base color tone (yellow and green) of PA 6,6 fabric surface, which is replaced with opposite coordinates in color wheel.

6.2.1.3 Spectrophotometric color combination of TVR-TPU-DMF-PA-6,6

Complex application of CM-3 states reddish (a^* = 22.45) and bluish (b^* = −13.30) hue on PA-6,6 fabric. A maximum value of L^* = 22.22 signifies the actual violet color of TVR on the surface of fabric. TVR added CM-03 accelerates the reddish and bluish hue.

6.2.1.4 Spectrophotometric color combination CBP-TVR-TPU-DMF-PA-6,6

CBP added CM-04 indicates minor decreasing the R/color hue in reddish and bluish hue although it seems the combination of CBP-TVR was not properly dispersed on the surface due to absence of hydrophilic vehicle, NA. Spectrophotometric color combination shows reddish (a^* = 13.06) and bluish (b^* = −12.95) hue on fabric surface. A maximum value of L^* = 32.34 indicated the hiding tendency of reddish and bluish hue on the surface of fabric when CBP was mixed in CM. Even though proper reaction of complex formation was not found due to absence of NA vehicle.

6.2.1.5 Spectrophotometric color combination NA-TVR-TPU-DMF-PA-6,6

NA added CM-05 shows that reflection of color hue increases. Spectrophotometric color combination of NA-TVR-TPU-DMF depicts reddish (a^* = 24.90) and yellowish (b^* = 26.75) hue. A maximum value of L^* = 28.2 indicates the increasing of reddish and bluish hue when NA was mixed in the CM. This is clarified that dyes molecules showed higher affinity with PA-6,6 when NA was mixed with TVR dyes.

6.2.1.6 Spectrophotometric color combination of NA-CBP-TVR-TPU-DMF-PA-6,6

CBP added CM-06 may be concealed the red and yellow hue of PA-6,6 surface when gray color wheel tends to black color hue. It seems that NA is functioning as hydrophilic transformation of CBP-TVR into PA-6,6. Combination of CBP mixture shows more absorbency of light as per less value of L^*. CBP modified PA-6,6 may be influenced the illumination of reflection of light/color hue. CBP modified PA-6,6 has an impact of camouflage creation and hiding factor of reddish and yellowish color under practical observation of CIE chromaticity-reflection mechanism. CBP may be acted as diminish behavior of reflection on PA-6,6 fabric surface which can create a camouflage effect of fabric under consideration CIE, L^*, a^*, b^*. It can be said that NA increased the thixotropic properties of CBP-TVR colorant for good level of CBP-TVR dispersion.

6.3 Camouflage and reflection comparison in NIR spectrum

Maximum reflection percentage of CM has been summarized in **Figure 17** for comparison among raw CBP, untreated and treated PA-6,6 fabric. The original FTIR spectra has been cited in supporting information, **Figures S1** and **S2**. FTIRS was used to reveal camouflage phenomenon of CBP in terms of low reflection principle in NIR, 1000–2500 nm. The range is covered by hyperspectral camera for target detection. CBP almost absorbs all spectrum in NIR, the intensity of color forming group is very minor for target identification. A narrow reflection profile is visualized for FTIRS scanning of raw CBP and CBP treated PA 6,6 fabric. CBP treated PA 6,6 fabric was detected as low reflection materials for camouflaging in NIR. CBP has maximum number of absorptions, and it has minimum intensity which may generate low reflection chromatic signal against combat background in NIR. Reflection of CBP spectra is always lower than any other combination of PA-6,6 fabric. Comparatively TVR combination cannot decline the reflectance of PA-6,6-TPU-DMF. CBP-TVR combination can highly decrease the reflection of PA-6,6-TPU-DMF. This concept of CBP-reflection can be implemented for synthetic dyes combination with CBP. It has been clearly signified that CBP treated fabric may act as target concealment under low reflection principle in NIR spectrum.

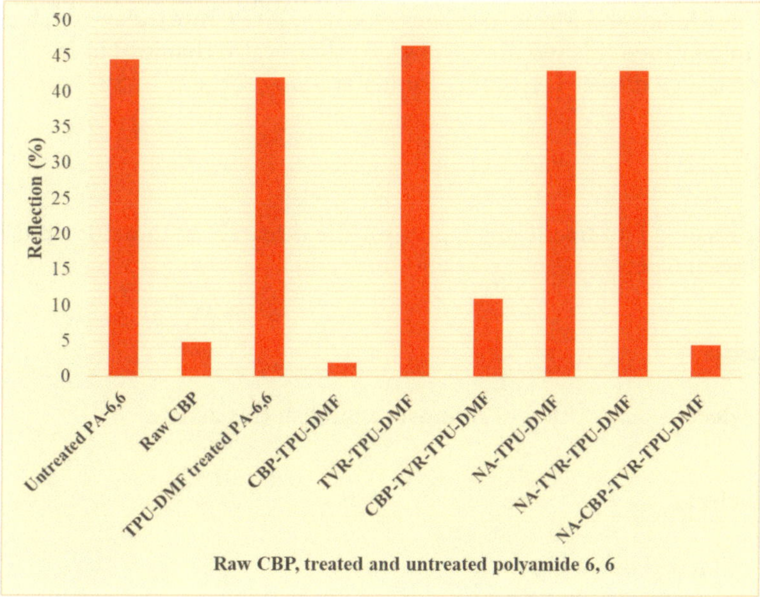

Figure 17.
Comparison of maximum reflection (%) in NIR spectrum.

7. Conclusion

L* coordinates reflect the lightness of color intensity to the observers ranges from 0 (completely black) to 100 (completely white). The luminous intensity and degree of lightness can be perceived by the target observers in remote sensing device. The amount of CBP adsorbed on PA-6,6 surface controls a* coordinates (red/green) and b* coordinates (yellow/blue), which identify the lightness of color perception for CBP modified PA-6,6 fabric to the observers. Hence, CBP influences a* coordinates for reflection of red or green hue to the target observers. Correspondingly CBP also effects on b* coordinates, which reflect yellow or blue color tone to the observers. The intensity of object reflection modifies the perception of red, green, blue, yellow color hue to the observers. Furthermore, CBP treated PA-6,6 fabric may generate diffuse reflection to hide the color hue of combat background. SEM, color measurement spectrophotometer and FTIRS have been confirmed the low reflection coloration of PA-6,6 fabric both in visible-NIR spectrum. The experimentation on CBP based camouflage textiles can confirm the feasibility of camouflage properties development on textile substances. So, it can be suggested that CBP can be implemented for camouflage coloration of textile materials in terms of low reflection principle. CBP can be combined with synthetic dyes of camouflage design for camouflage textiles in simultaneous spectrum probe in visible-NIR. CBP may be accepted for versatile applications of weapon/vehicle coloration in terms of low reflection principle of camouflage coloration. TPU can also be recommended as a suitable cross-linking agent due to transparent property of TPU without influencing the reflection properties of CBP on PA-6,6 fabric.

Acknowledgements

Author Md. Anowar Hossain acknowledges RMIT University & Australian Government for funding through RTP Stipend Scholarship. Author acknowledges to "Professor Lijing Wang" and "Emeritus Professor Robert Shanks," School of Fashion and Textiles, RMIT University for their draft review. Author is indebted to Dr. Olga Gavrilenko and Mr. Imtiaz Ahmed Khan, Ph.D. researcher for sharing their materials to continue this research works. Author is also thankful to Ms. Shelley MacRae, technical officer (research) for her training support of SEM and FTIRS.

Funding

Author received no financial support for the research, authorship, and/or publication of this article.

Declaration

Author declare no conflict of interest to publish this article.

Abbreviations

CBP	Carbon black pigment
CM	Composite mixture
NR	Near infrared
NA	Nitric acid

TVR	Telon violet 3R
TPU	Thermoplastic polyurethane
DMF	N, N-dimethylformamide
PA-6,6	Polyamide-6,6

Appendix

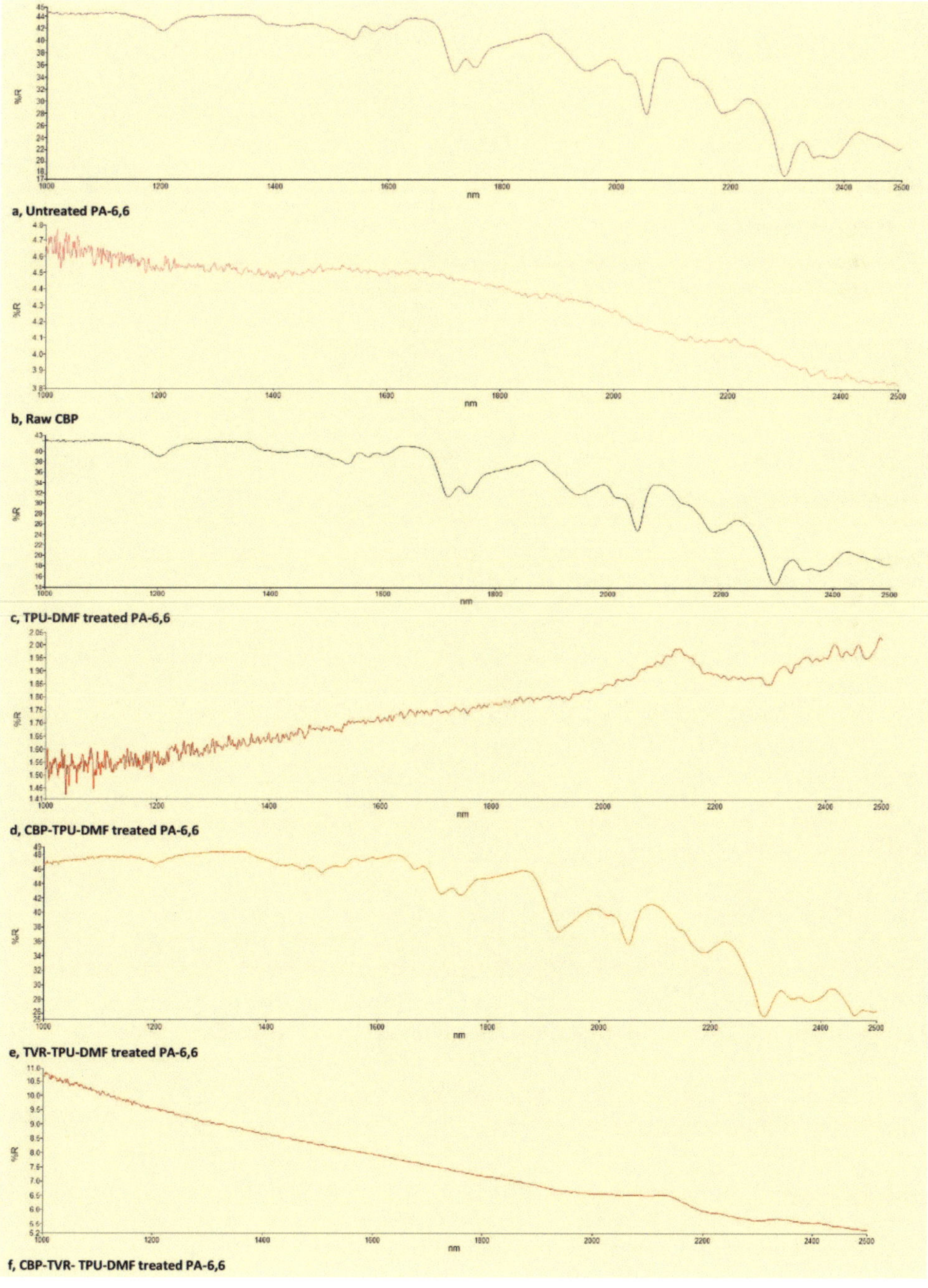

Figure S1.
Comparison of reflection (%) in NIR spectrum (part-a).

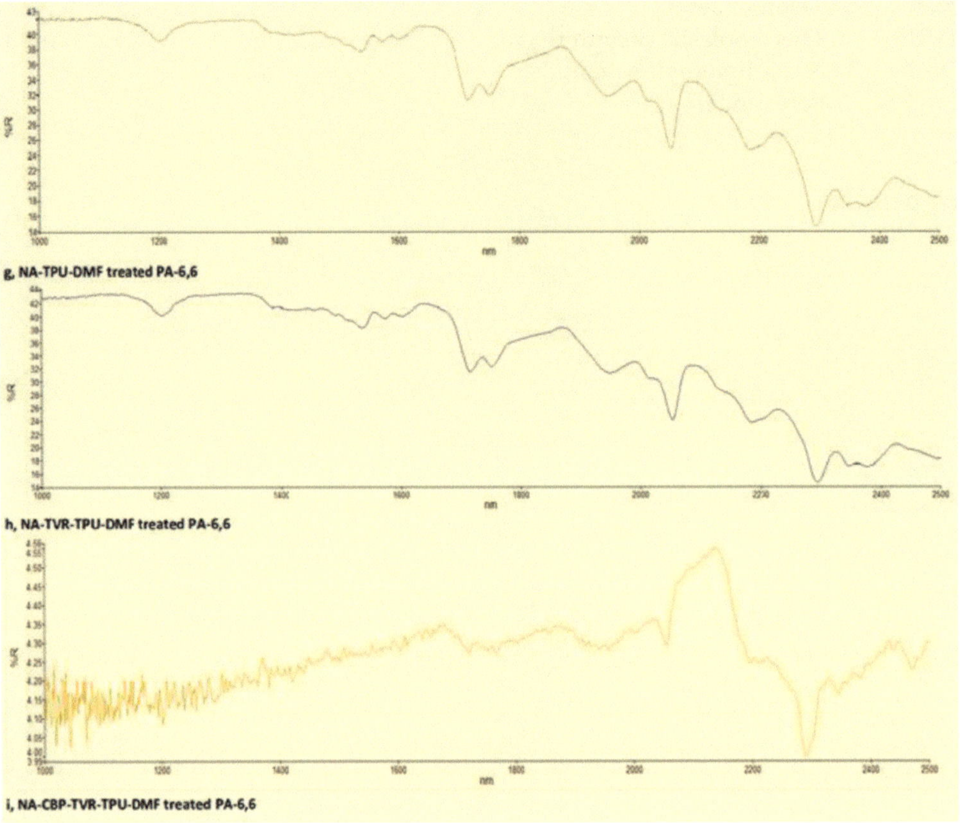

Figure S2.
Comparison of reflection (%) in NIR spectrum (part-B).

Author details

Md. Anowar Hossain[1,2]

1 School of Fashion and Textiles, RMIT University, Melbourne, VIC, Australia

2 Department of Textile Engineering, City University, Dhaka, Bangladesh

*Address all correspondence to: engr.anowar@yahoo.com; s3820066@student.rmit.edu.au

References

[1] Sinha MA. Defence Textiles & Camouflage Fabrics. Bangkok: International Conference-Global Textile-Opportunities & Challenges in an Integrated World, The Textile Association, India, Global Textile Congress; 2015

[2] Aiken D, Ramsey S, Mayo T. First-order Parametric Model of Reflectance Spectra for Dyed Fabrics. Washington, DC: Naval Research Laboratory; 2016. pp. 20375-25320

[3] Aiken D, Ramsey S, Mayo T. Parametric Models of NIR Transmission and Reflectivity Spectra for Dyed Fabrics. Vol. 20375-5320. Washington, DC: Naval Research Laboratory; 2015. pp. 1-33

[4] Clarkson GM. Camouflage Fabric. USA: U.S. Patent, Editor; 1998

[5] Rudolf W. Process for the production of camouflage dyeings and prints. Hoechst Aktiergesellschaft, Frankfurt an Main, Germany: USA: W. Patent Office, Editor; 1978

[6] Vitalija R et al. Development of visible and near infrared camouflage textile materials. Materials Science (MEDŽIAGOTYRA). 2009;**15**(2): 173-177

[7] Belmont JA, Boes RU, Menashi J. Carbon Black Products for Coloring Mineral Binders, U.S. Patent, Editor. Boston, Mass, USA: Cabot Corporation; 1996

[8] Moharam ZE, Tavanaie MA, Mehrizi MK. Reflectance properties of brown mass dyed poly(ethylene terephthalate) filament yarns in the visible-near infrared region. Progress in Color, Colorants and Coatings. 2020;**13**: 93-104

[9] Omae Y et al. Carbon Black Aggregate, U.S. Patent, Editor. Tokyo, Japan, USA: Mitsubishi Chemical Corporation; 2000

[10] Geary CG, Amboy P. Ceramic Coloring Agent, U.S. Patent, Editor. Wilming ton, Del, a corporation of Delaware: USA: E. I. duPont de Nemours & Company; 1938

[11] Mori Y et al., editors. Carbon black-graft polymer, method for production thereof, and use there of, U.S. Patent, Editor. Osaka, Japan, USA: Nippon Shokubai Kagaku Kogyo Co., Ltd; 1991

[12] Ikeda H, Urashima N, Ando N. Carbon Black Graft Polymer, U.S. Patent, Editor. Osaka (JP), USA: Nippon Shokubai Co., Ltd; 2002

[13] Weihe A. Process of Modifying Carbon Black, U.S. Patent, Editor. a corporation of the Federal Republic of Germany: USA: Deutsche Gold- und. Silber-Scheideanstalt Wormals Roessler, Frankfurt am Main, Germany; 1957

[14] Chaplin TR et al. Method for Forming an Aqueous Carbon Black Dispersion, U.S. Patent, Editor. Philadelphia, PA (US), USA: Rohm and Haas Company; 2006

[15] Donnet J-B. Fifty years of research and progress on carbon black. Carbon. 1994;**32**(7):1305-1310

[16] Swart GH. Akron, and Ohio, Carbon Black Pellets, U.S. Patent, Editor. Akron, Ohio, a corporation of Ohio: USA: The General Tire & Rubber Company; 1947

[17] O'Neill WO. Printing Inks Containing Styrene Maleic Anhydride Copolymer and Carbon Black, U.S.P. Office, Editor. Rochester, N.Y., a corporation of New Jersey: USA: East man Kodak Company; 1962

[18] Wildeman GF, Virant AJ. Pigment Dispersions Exhibiting Improved

Compatibility in Polyurethane Systems. USA: U.S.O. Patent, Editor; 1988

[19] Suetsugu K, Maruyama E, Same K. Black Coloring Agent, U.S. Patent, Editor. Tokyo, Japan, USA: Mitsubishi Chemical Industries, Ltd; 1976

[20] Nguyen VT, Bellmann G, Boussel F. Low Viscosity Stable Aqueous Dispersion of Graft Carbon Black, U.S. Patent, Editor. Geneva, Switzerland, USA: Battelle Memorial Institute; 1985

[21] Graf H, Riffel D. Wetting Agent-Containing Pigment Composition, U.S. Patent, Editor. Germany, USA: Deutsche Gold-und Silber-Scheideanstalt Vormals Roessler, Frankfurt am Main; 1972

[22] Moriya S. Three-Dimensional Coloring Composition. USA: U.S. Patent, Editor; 1969

[23] Eisenmenger E et al. Carbon Black Useful for Pigment for Black Lacquers, U.S. Patent, Editor. Fed. Rep. of Germany: USA: Deutsche Gold- und Silberscheideanstalt Wormals Roessler; 1982

[24] Ferch, H., H. Wagner, and Maintal, Carbon Black Preparation for Use in Mineral Binder, U.S. Patent, Editor. 1977, Deutsche Gold- und Silber-Scheideanstalt vormals Roessler, Frankfurt, Germany, USA

[25] Bowden AP. Cement Coloring Composition and Method of Producing Same, U.S. Patent, Editor. King of Prussia, Pa.: USA: M. Hamburger & Sons, Inc.; 1980

[26] Augustin F, Bartling L. Packed Pigment Pastes and Process For Coloring and Toning Water-Diluted Coating Agents, U.S. Patent, Editor. Fed. Rep. of Germany, USA: BASF Farben & Fasern AG, Hamburg; 1980

[27] Fitch WL, Everhart ET, Smith DH. Characterization of carbon black adsorbates and artifacts formed during extraction. Analytical Chemistry. 1978; **50**(14):2122-2126

[28] Suga Y, Shimomura M. Inkjet Recording Having an Ink With Carbon Black, U.S. Patent, Editor. Tokyo, Japan, USA: Canon Kabushiki Kaisha; 1993

[29] Hirosawa FN, Lee MH. Polyurethane Cross-Linking Agents, U. S. Patent, Editor. Los Angeles, Calif, USA: Furane Plastics, Inc; 1977

[30] Kantouch FA, El-Sayed AA. Acid dyeable and printable acrylic fabrics treated with cationic aqueous polyurethane. Journal of Applied Polymer Science. 2011;**119**:2595-2601

[31] Elwakil HA. Encapsulated Magnetic Particles Pigments and Carbon Black, Compositions and Methods Related There To, U.S. Patent, Editor. Wood Dale, Ill: USA: Videojet Systems International, Inc; 1997

[32] Kothandaraman H, Venkatarao K, Thanoo BC. Preparation, properties, and crosslinking studies on polyurethane elastomers. Polymer Journal. 1989; **21**(10):829-839

[33] Bamfield P, Hutchings M. Chromic Phenomena, Technological Applications of Color Chemistry. Royal Society of Chemistry; 2008. p. 459

[34] Rodgers B, Waddell W. Tire engineering. In: Mark JE, Erman B, Roland CM, editors. Science and Technology of Rubber. The Boulevard, Oxford, UK: Academic Press, MA, USA; 2013. p. 644

[35] Swinehart DF. The Beer-Lambert law. Journal of Chemical Education. 1962;**39**(7):333-335